Textbook of Fiber Formation Technology

By Qing Shen
沈青 编著

纤维成型技术教程

Donghua University Press
東華大學出版社
·上海·

内容提要

《纤维成型技术教程》是一本以英文编写的教材，在每一章的结束附有相应的习题，主要面向高分子材料专业的本科生、研究生，也适合其他读者阅读参考。

本书的内容分成7个部分，第1章介绍纤维的历史；第2章介绍纤维的定义、结构和分类；第3章介绍16种根据应用命名的纤维，如工业纤维、复杂纤维、军用纤维、智能纤维等等；第4章介绍主要的纤维及其来源，如天然纤维、合成纤维等；第5章主要介绍23种纺丝技术，如熔融纺丝、湿法纺丝、干法纺丝、凝胶纺丝等；第6章主要介绍伯测试与表征技术；第7章主要通过实例介绍不同的纤维成型技术，涉及11个例子，如一步法合成技术直接制备纳米纤维等。

图书在版编目(CIP)数据

纤维成型技术教程/沈青编著. —上海：东华大学出版社，2015.6

ISBN 978-7-5669-0779-0

Ⅰ.①纤… Ⅱ.①沈… Ⅲ.①化学纤维—成型—工艺学—教材 Ⅳ.①TQ340.6

中国版本图书馆CIP数据核字(2015)第097497号

责任编辑：张 静

封面设计：魏依东

出 版：东华大学出版社(上海市延安西路1882号，200051)

本社网址：http://www.dhupress.net

淘宝书店：http://dhupress.taobao.com

营销中心：021-62193056 62373056 62379558

印 刷：崇明裕安印刷厂

开 本：787 mm×1 092 mm 1/16 印张 10.5

字 数：282千字

版 次：2015年6月第1版

印 次：2015年6月第1次印刷

书 号：ISBN 978-7-5669-0779-0/TS·604

定 价：29.00元

Preface

Fiber plays a key role in many areas and the past century has witnessed a huge increase of research interest in the study of fiber products and quantity development. Now, the properties of various fibers in synthetic, chemical and natural all are very importantly applied elsewhere.

This book provides a brief introduction of fiber knowledge and various fiber technologies as well as related cases on fiber formation. The fiber types represent their applications which we met in life and somewhere covering main aspects of cutting edge in research and development.

This book is organized by seven chapters with total not more pages to fit the course study for students in China. Chapter 1 introduced the fiber history. Chapter 2 consists of five parts and deals with fundamental aspects of fiber knowledge such as fiber definition, structure and classification. Chapter 3 described various fiber types in relation to their possible application. Chapter 4 contains main fiber resources, e.g. from natural and synthetic polymers or from inorganic materials, respectively. Chapter 5 introduced main fiber formation techniques and related equipment, where the melt spinning was in detail described. Chapter 6 deals with main methods for fiber measurement and characterization where the fiber morphology, structure, properties and surface wetting were introduced. Chapter 7 presented several detailed cases on fiber formation which was considered to help students to direct understand fiber formation.

Of this textbook, the recommended reading was arranged in each chapter and this was expected to fit the further reading for students. In some chapters, problems were also appeared in the final to lead students to think some novel question on development of fiber techniques. Also, it is greatly wished that this textbook can help students to start advanced study internationally.

I would like to thank my family for their friendly and courteous assistance.

Qing Shen

Contents | 目录

Chapter 1

Fiber History

Fiber is defined as a solid material with stable thin shape and long size as well as certain level of tensile strength. As has been consistently recognized that the fiber science and technology were learnt from silkworm and spider since their fiber producing processes are good examples of the biosynthetic and biospinning techniques because they convert non-fiber foods by enzymes into proteins in body then spun fibers as a cocoon or net. In addition to animal fibers, plants-based fibers are generally synthesized in natural conditions, e.g. by photo, carbon dioxide and water.

The use of fiber in human history can back to the Paleolithic times by observing the ancient people used cordage in fishing, trapping and transport, as well as the fabrics for clothing using palm leaf, jute, flax, ramie, sedges, rushes and reeds as raw materials. The first true paper is made in southeastern China in the second century AD from old rags, hemp and ramie and later from the fiber of the mulberry tree.

For thousands of years, the fiber use was limited by the inherent qualities available in the natural world, e. g. cotton and linen wrinkled from wear and washings, silk required delicate handling, and wool shrank irritating to the touch and eaten by moths. A mere century ago, the first manufactured fiber, rayon, was developed and soon applied to people's life. The secrets of fiber chemistry for countless applications must be emerged.

Man manufactured fibers now are put to work in modern apparel, home furnishings, medicine, aeronautics, energy and various industry areas, and these lead fiber engineers to combine or modify fibers in some ways far beyond the performance limits of fiber drawn from the silkworm cocoon, grown in the fields, or spun from the fleece of animals.

The earliest published record of an attempt to create an artificial fiber took place in 1664 when the English naturalist *Robert Hooke* suggested the possibility of producing a fiber that would be "if not fully as good, nay better" as compared with the silk.

The first patent for "artificial silk" was granted in England in 1855 contributed by a Swiss chemist, *Audemars*. He dissolved the fibrous inner bark of a mulberry tree; to chemical modify it to produce cellulose. He formed threads by dipping needles into this solution and drawing them out, however, it never lead him to emulate the silkworm by extruding the cellulosic liquid through a small hole.

In the early 1880's, an English chemist and electrician, Sir *Joseph W. Swan*, was spurred to action by *Thomas Edison*'s new incandescent electric lamp. He experimented with forcing a liquid similar to *Audemars's* solution through fine holes into a coagulating bath, obtained fibers worked like carbon filament.

It also occurred to *Swan* that his filament could be used to make textiles. In 1885 he exhibited some fabrics crocheted by his wife from his new fiber in London.

The first commercial scale production of manufactured fiber was achieved by French chemist, *Count Hilaire de Chardonnet*, in 1889, when his fabrics of "artificial silk" caused a sensation at the Paris Exhibition. After that about two years, he built the first commercial rayon plant at Besancon, France. On the basis of this great contribution, he has fame as the "father of the rayon industry".

In United States, several attempts have been made to produce "artificial silk" during the early 1900's but none were commercially successful until the American Viscose Company formed by Samuel Courtaulds Co., Ltd. to produce rayon in 1910.

In 1893, *Arthur D. Little* of Boston, invented another cellulosic product, acetate, and developed it as a film. By 1910, *Camille* and *Henry Dreyfus* made acetate motion picture film and toilet articles in Basel, Switzerland. During the World War Ⅰ, they built a plant in England to produce cellulose acetate dope for airplane wings and other commercial products. Upon entering the War, the United States government invited the Dreyfus brothers to build a plant in Maryland to make the product for American warplanes. The first commercial textile uses for acetate in fiber form were developed by the Celanese Company in 1924.

In the meantime, U.S. rayon production was growing to meet increasing request, and by the mid of 1920's, textile manufacturers could purchased the fiber for half the price of raw silk.

In 1931, American chemist, *Wallace Carothers*, reported their research results carried out in the laboratory of the DuPont Company on "giant" molecules called polymers. He focused his work on a fiber referred to simply as "66", a number derived from its molecular structure. Nylon, the "miracle fiber", was born. The Chemical Heritage Foundation is currently featuring an exhibit on the history of nylon.

In 1938, a scientist, *Paul Schlack*, worked at the I. G. Farben Company in Germany, polymerized the caprolactam and created a different form of the polymer, which was identified simply as nylon "6".

Nylon's coming created a revolution in the fiber industry because rayon and acetate have been derived from plant cellulose, while nylon was synthesized completely from petrochemicals to establish the basis for the ensuing discovery of an entire new world of manufactured fibers *via* human synthesis.

DuPont began commercial producing nylon in 1939, when the first experimental testing used nylon as sewing thread, in parachute fabric, and in women's hosiery. Nylon stockings were shown in February 1939 at the San Francisco Exposition, and since then most excited fashion innovations of the age were underway.

During the World War Ⅱ, nylon replaced Asian silk in parachutes and used in tires, tents, ropes, ponchos, and other military supplies, and even was used in the production of a high-grade paper for U.S. currency. At the outset of the War, cotton was king of fibers, accounting for more than 80% of all fibers used. Manufactured and wool fibers shared the remaining 20%. By the end of the War in August 1945, cotton stood at 75% of the fiber market. Manufactured fibers had risen to 15%.

After the War, GI's came home, families were reunited, industrial America gathered its peacetime forces, and economic growth surged. The conversion of nylon production to civilian uses started and when the first small quantities of postwar nylon stockings were advertised, leading thousands of frenzied women lined up at New York department stores to buy.

In the immediate post-war period, most nylon production was used to satisfy this enormous pent up demand for hosiery. But by the end of the 1940's, it was also being used in carpeting and automobile upholstery. At the same time, three new generic manufactured fibers started production at Dow Badische Company (today, BASF Corporation), e. g. the metalized fibers; at Union Carbide Corporation, e. g. the modacrylic fiber; and at Hercules, Inc. e. g. the olefin fiber. Since that period, the man-made fibers continued their steady march.

By the 1950's, the industry was supplying more than 20% of the fiber needs of textile mills. A new wool-like fiber, "acrylic", was developed at DuPont.

Meanwhile, polyester, first examined as part of the Wallace Carothers in early researches, was attracting new interest at the Calico Printers Association in Great Britain. There, *J. T. Dickson* and *J. R. Whinfield* produced a polyester fiber by condensation polymerization of ethylene glycol with terephthalic acid. DuPont subsequently acquired the patent rights for the United States and Imperial Chemical Industries for the rest of the world. A host of other producers soon joined in.

In the summer of 1952, "wash and wear" was coined to describe a new blend of cotton and acrylic, and this term was eventually applied to a wide variety of manufactured fiber blends including the polyester fiber leading a revolution in textile product performance.

Polyester was commercialized in 1953 by the introduction of triacetate. In the 1960's and 1970's, consumers bought more and more clothing made with polyester. Clotheslines were replaced by electric dryers, and the "wash and wear" garments they dried emerged wrinkle free. Ironing began to shrink away on the daily list of household chores. Fabrics became more durable and color more permanent. New dyeing effects were being achieved and shape-retaining knits offered new comfort and style.

In fact, in the 1960's, the manufactured fiber production was accelerated by continuous fiber innovation. The revolutionary new fibers were modified to offer greater comfort, provide flame resistance, reduce clinging, release soil, achieve greater whiteness, special dullness or luster, easier dyeability, and better blending qualities. New fiber shapes and thicknesses were introduced to meet special needs. Of which, the Spandex, a stretchable fiber; aramid, a high-temperature-resistant polyamide; and para-aramid, with outstanding strength-to-weight properties, were introduced into the marketplace.

One dramatic new set of uses for manufactured fibers came with the establishment of the U. S. space program. The industry provided special fiber for uses ranging from clothing for the astronauts to spaceship nose cones. When Neil Armstrong took "One small step for man, one giant leap for mankind", on the moon on July 20, 1969, his lunar space suit included multi-layers of nylon and aramid fabrics. The flag he planted was made of nylon.

Today, the exhaust nozzles of the two large booster rockets that lift the space shuttle into orbit contain 30 000 pounds of carbonized rayon. Carbon fiber composites are broadly used as structural components in the latest commercial aircraft, adding strength and lowering weight and fuel costs.

Innovation is the hallmark of the manufactured fiber industry. More numerous and diverse fibers than any found in nature are now routinely created in the laboratories of industry.

Nylon variants, polyester, and olefin are used to produce carpets that easily can be rinsed clean-even 24 hours after they've been stained. Stretchable Spandex and machine-washable, silk-like polyesters occupy solid places in the U.S. apparel market. The finest microfibers are remaking the world of fashion.

For industrial uses, manufactured fibers relentlessly replace traditional materials in applications from super-absorbent diapers, to artificial organs, to construction materials for moon-based space stations. Engineered non-woven products of manufactured fibers are found in applications from surgical gowns and apparel interfacing to roofing materials, road bed stabilizers, and floppy disk envelopes and liners. Non-woven fabrics, stiff as paper or as soft and comfortable as limp cloth, are made without knitting or weaving.

Below we listed the innovation history for each man-made fiber. The first man-made cellulose fiber was reported by *Nicolaus de Chardonnet* in 1884. The first cooper silk filaments was reported in 1898 in Oberbruch near Aachen of Germany, by *Paul Fremery*, *Bromert* and *Urban*. The first polyvinyl chloride, PVC, was produced by *Klatte* in 1913. In 1927 *Staudinger* first spun the fully synthetic fiber from polyoxymethylene and later from polyethylenoxide by melting spinning. In 1934, the first semitechnical production of polyaerylonitrile fibers (PAN) was reported in Germany. The polyurethane, PU, fiber was developed in 1937 by *O. Bayer* et al. In

1938, *Carothers* reported the first polycondensation fiber, Nylon, in DuPont de Nemours & Co. Following, *Schlack* produced the lactam-based fiber, Perlon®, in 1939 in Berlin-Lichtenberg, Germany. The commercial polyester fibers were reported around 1950 developed by *Whinefield*. The first polypropylene, PP, fiber was reported around 1958 developed by *NaUa*. The high performance fiber, e.g. Nomex® and Kevlar®, both were developed by DuPont in 1963 and 1970, respectively. The development of high-grade carbon fibers was in 1966 with the oxidation and carbonization of PAN filaments. The novel biodegradable fiber, poly(lactic acid), PLA, was developed by Cargill Dow Chemical Company in 1997.

Among above periods, a lot of developed fibers without commercialized, e. g., polyaminotriazoles fiber, polyamides 4 and 7, PA 4 and PA 7, due to spinning difficult or cost reasons.

Recommending reading

[1] M. Harris, ed.. *Handbook of Textile Fibers*. Harris Research Laboratories, Inc., Washington, D. C., 1954.

[2] J. G. Cook. *Handbook of Textile Fibers*. *Vol. I: Natural Fibers*. 5th ed.. Merrow Publishing, Durham, N.C., 1984.

[3] T. Hongu, G. O. Phillips, M. Takigami. *New Millennium Fibers*. CRC Press, Woodhead Publishing Limited, 2005.

Chapter 2

Fiber Definition, Structure and Classification

2.1 Fiber definition

Fiber is defined as a solid material with stable thin shape and long size as well as certain level of tensile strength.

2.2 Fiber morphology and structure

The morphologies of some fibers are showed in Figure 2-1. It is clearly that the plant-based fibers, e.g. flax and cotton, with pore structure, while the animal-based fibers, e.g. wool and silk, having solid structure.

The fiber structure is generally known in micro- and macro-state as Figures 2-2 and 2-3 described, respectively.

Basically, it was known that the fiber consists of three distinct phases, i.e. the oriented crystalline regions, the amorphous regions also with preferential orientation along the fiber axis which contain tie molecules connecting crystallites, and the highly extended non-crystalline molecules which was called the interfibrillar phase. In these three phases, the interfibrillar phase plays a key role in the tensile properties of the fiber.

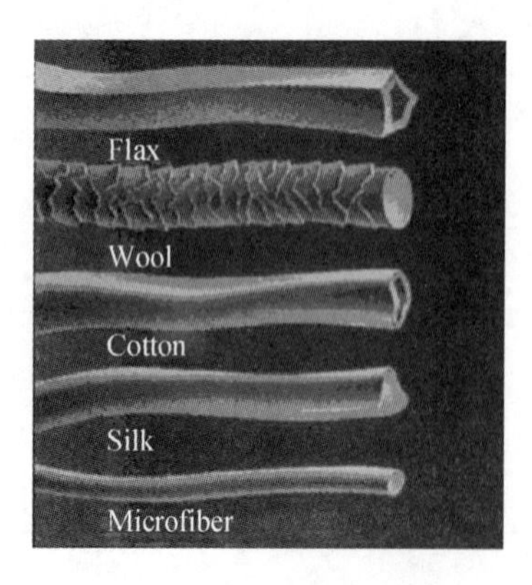

Figure 2-1 Morphologies of several fibers

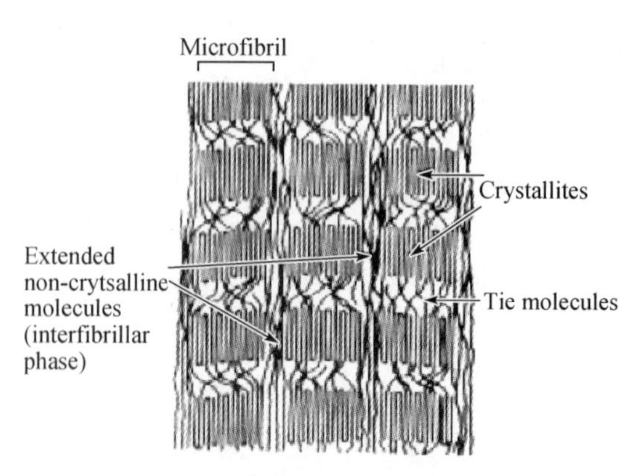

Figure 2-2 A scheme of the micro structure of fiber

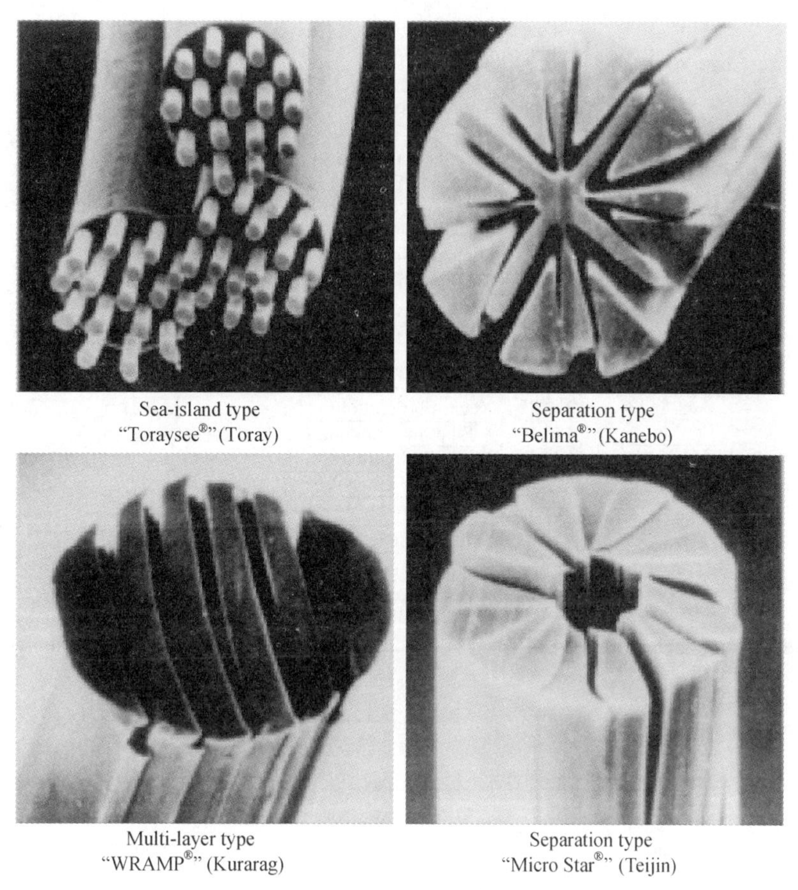
Sea-island type
"Toraysee®"(Toray)

Separation type
"Belima®"(Kanebo)

Multi-layer type
"WRAMP®" (Kurarag)

Separation type
"Micro Star®" (Teijin)

Figure 2-3 The macro-structure of fiber

2.3 Fiber classification

The classification of fiber has been defined by many people while yet without a standard agreed with all scientists and engineers.

Man-made fibers are probably classified into three classes according to the raw materials, e. g. the natural polymeric fiber, the synthetic polymeric fiber and the inorganic material-based fiber.

2.4 Fiber diameter and cross-sections

The fiber diameter-based classification can be understood from Figure 2-4, where the material kind related information appeared correspondingly. Of which it must be addressed that the nanofibers are also produced available by people using different methods, and here presented examples only for those in their initial size.

The common shape of fibers defined according to the cross-section morphology was described in below (Figure 2-5).

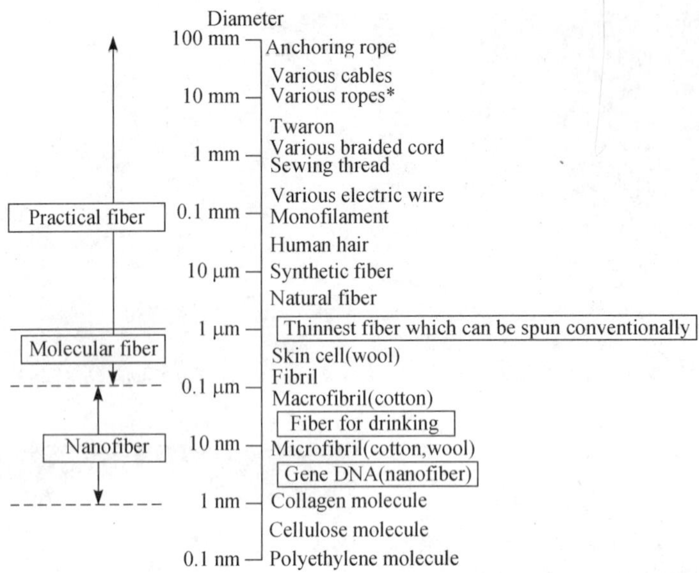

*: Distinction of fiber and assembly of fibers is necessary(Possible to make rope thicker)

Figure 2-4　A classification of fiber based on diameter in relation to different materials

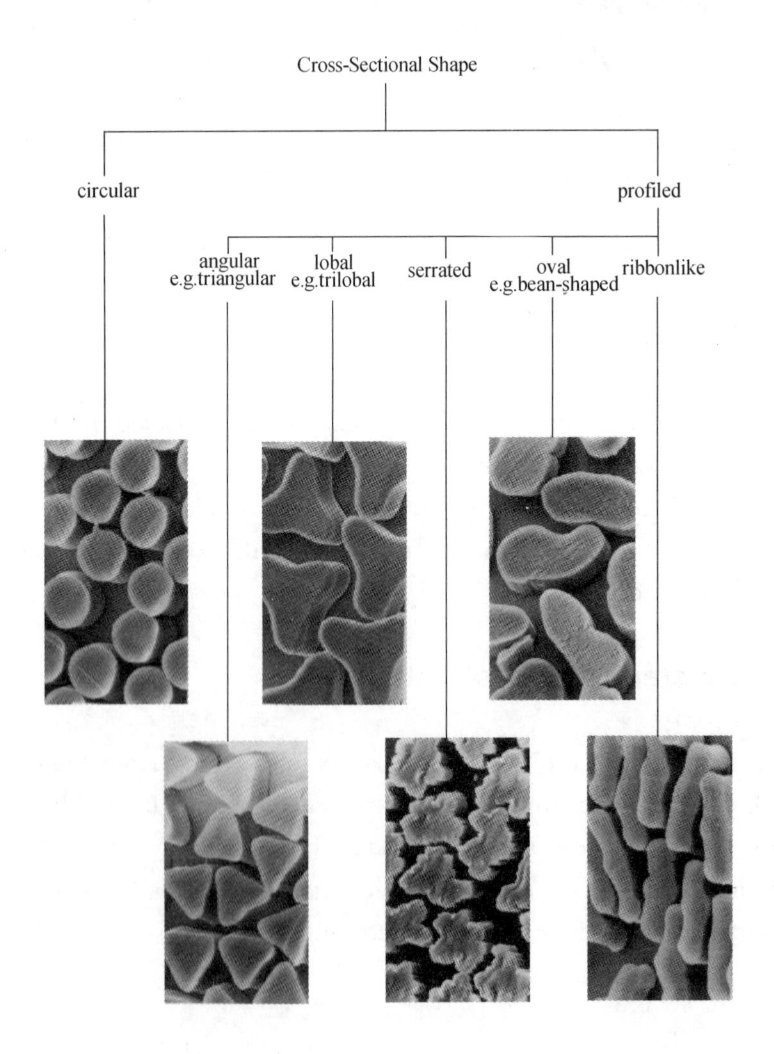

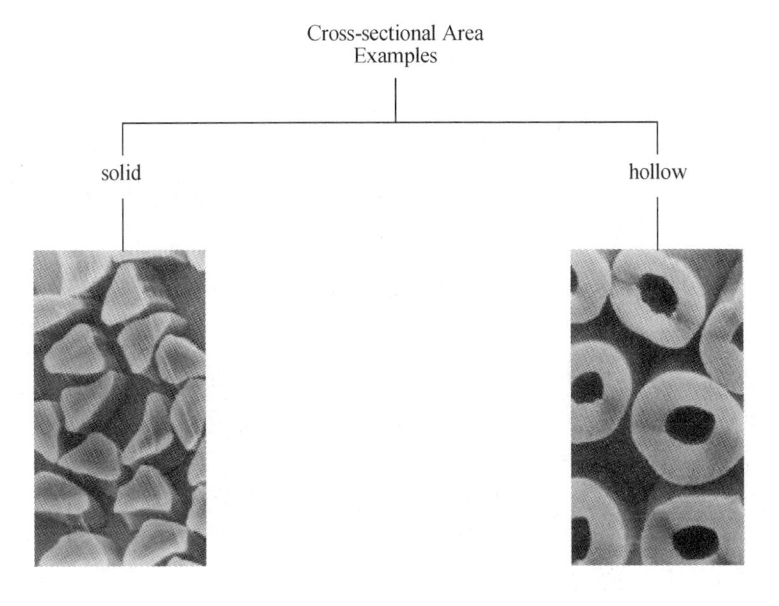

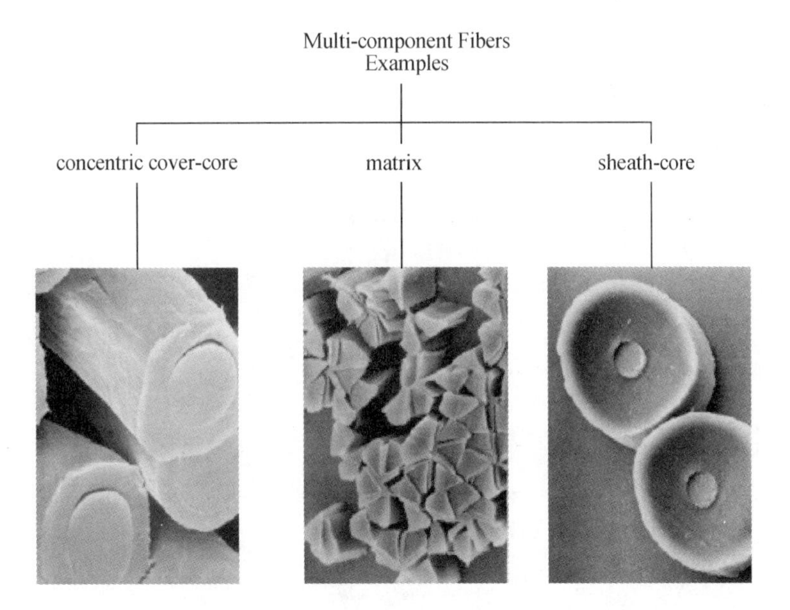

Figure 2-5 Cross-section of several fibers (Franz Fourne. *Synthetic Fibers, Machines and Equipment, Manufactory, Properties.* Hanser Publisher, Germany, 1998)

2.5 Importance of fiber technology

Below is a summarization of the fiber importance on the basis of the application.

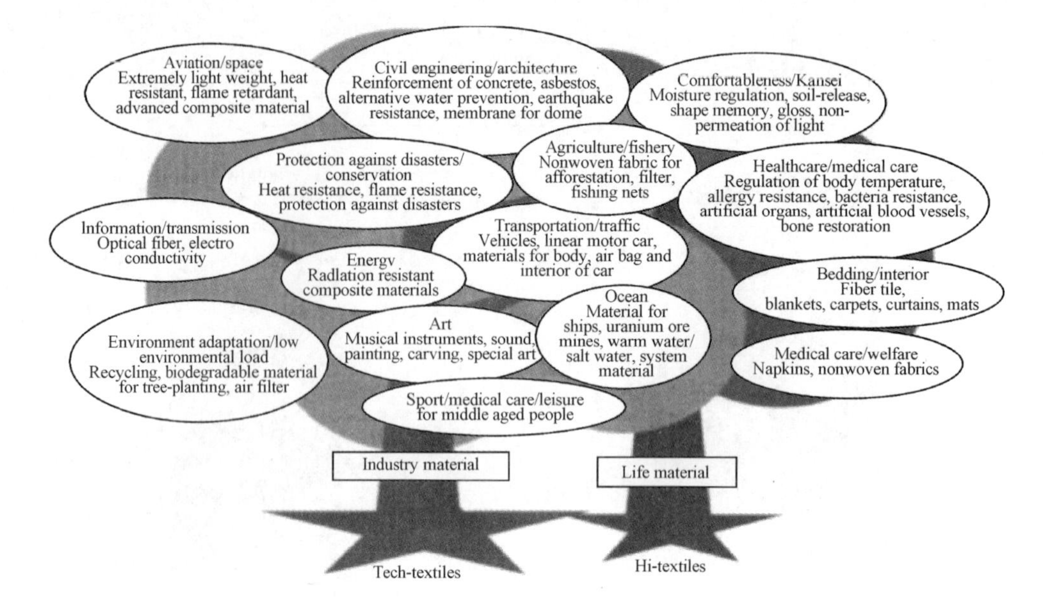

Figure 2-6　A summary of the use of fibers in today (T. Hongu, G. O. Phillips, M. Takigami. *New Millennium Fibers*. CRC Press, Woodhead Publishing Limited, 2005)

Recommending reading

[1] M. Harris, ed.. *Handbook of Textile Fibers*. Harris Research Laboratories, Inc., Washington, D. C., 1954.

[2] J. G. Cook. *Handbook of Textile Fibers*. 5th ed.. Merrow Publishing, Durham, N.C., 1984.

Problems

1. In terms of the fiber definition, does the steel needle or steel tube can be defined as a fiber respectively?

Chapter 3

Fiber Types

3.1 Industrial fiber

The classification of fiber on the basis of its use has been broadly applied and the industrial fiber is used for describing the fibers currently broadly applied in different industrial areas. However, this fiber is usually for those currently applied in car, road and related industries.

3.2 Textile fiber

The fibers used for textile industry are defined as textile fiber and these fibers are usually those made from cellulose and some synthetic polymers, e.g. PET, PAN, etc.

3.3 Military fiber

All those fibers would be applied for army products then probably be defined as the military fibers.

3.4 Synthetic fiber

The fiber made using the petrochemicals as raw materials was defined as the synthetic fibers, e.g. the polyester fiber.

3.5 Natural fiber

Fibers classified as natural fibers are vegetable, animal, or mineral in origin. As the name implied, the vegetable fibers are usually derived from plants and the principal chemical components are cellulose, which has been also referred to the cellulosic fibers and which are basically bound to another natural phenolic polymer, lignin. With respect to these two main natural polymers, the vegetable fibers therefore are also referred to as the lignocellulosic fibers, except the cotton, which contain few lignin.

The vegetable fibers are classified according to their source in plants as the bast or stem fibers, which are often used as the soft fibers for textile use; the leaf fibers, which are referred to as the hard fibers; and the seed-hair fibers, which is the most important vegetable fiber.

3.6 Organic fiber

In general, these fibers are the same as those synthetic fibers using petrochemical products as raw materials.

3.7 Inorganic fiber

These fibers are made using inorganic masses as raw materials, e.g. the silicon fiber (Figure 3-1).

Figure 3-1 Photograph of silicon fiber

3.8 Smart fiber

These fibers are of new and generally defined based on their special properties such as the condition-responses. For example, some fibers have the temperature, time and pH responses.

3.9 High performance fiber

The high performance fibers are generally characterized by they presented remarkably high unit tensile strength and modulus as well as resistance to heat, flame, and chemical agents. The high performance fibers are usually contributed by rigid-rod polymers due to their liquid crystalline state, e.g. classified as lyotropic, such as the aramid Kevlar (DuPont), or thermotropic liquid crystalline polymers, such as Vectran (Celanese).

The applications of these fibers are usually in the aerospace, biomedical, civil engineering, construction, protective apparel, geotextiles and electronic areas.

Poly(1, 4-benzamide), PBA, might be the first reported nonpeptide synthetic polymer to form a liquid crystalline solution. To obtain a liquid-crystalline solutions of poly(1, 4-benzamide), the first step is to prepare the polymer in a proper solvent, e.g. *N*, *N*-dialkylamide, *N*, *N*-dimethylacetamide, and *N*, *N*, *N*, *N*-tetramethylurea. To obtain the high molecular weight, the lithium base, e.g. lithium hydride, lithium carbonate, or lithium hydroxide, is usually adopted in the polymerization solution during the reaction time posted 1 ～ 2 h. Thereafter, thermotropic polyesters, e.g. the acidolysis caused poly (ethylene terephthalate), PET, by *p*-acetoxybenzoic acid, copolymer compositions that contained 40～70 mol% of the oxybenzoyl unit formed anisotropic, turbid melts which were easily oriented. Polyesters such as poly (*p*-phenylene terephthalate), which can form liquid crystalline phases, decompose at temperatures below the melting point.

The first reported high performance fiber is the polyaramid fiber, which was made based on poly (*m*-phenylene isophthalamide). Though this fiber was not liquid crystalline, the commercialized name by DuPont as Nomex nylon in 1963 and changed to Nomex aramid in 1972.

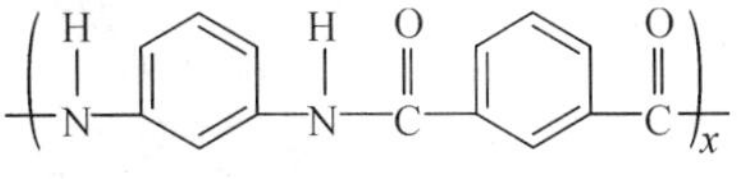

Nomex

In 1970s, Kevlar fiber was developed by DuPont by processing of extended chain all para-aromatic polyamides from liquid crystalline solutions produced ultrahigh strength, ultrahigh modulus. The greatly increased order and the long relaxation times in the liquid crystalline state compared to conventional systems led this fiber having highly oriented domains of polymer molecules.

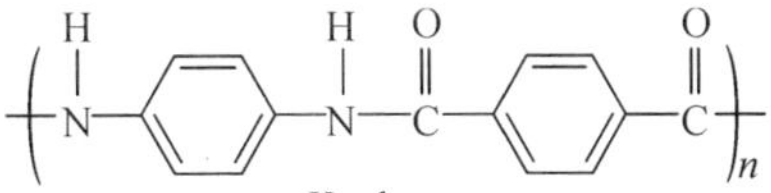

Kevlar

3.10 Special use fiber

Generally, this definition is not accurately, because all fibers are possibly to fit this definition. However, it was found that the special use seems to be for fibers used in medical, military, aerospace etc.

3.11 Super fiber

Generally, the super fibers represent some fibers with remarkable mechanical properties, e.g. the strength more than *ca*. 2 GPa, and the elastic constant more than *ca*. 50 GPa. Notable, these units are absolutely different than the traditional fibers due to the latter represented in units of cN/dtex (centi-Newton/deci-tex). The value in cN/dtex represents the load per unit line density. On the other hand, the value in GPa represents the load per unit sectional area and is larger than that in cN/dtex. For example, the Para-aramid superfiber has a strength of 20 cN/dtex and elastic constant 500 cN/dtex (density = *ca*. 1.44 g/cm^3) corresponds to 2.9 GPa and 72 GPa, respectively.

3.12 Dietary fiber

Obviously, these fibers are eatable, e. g. some protein-based natural fibers. The fiber from lotus root is a good example of dietary fiber as showned in below.

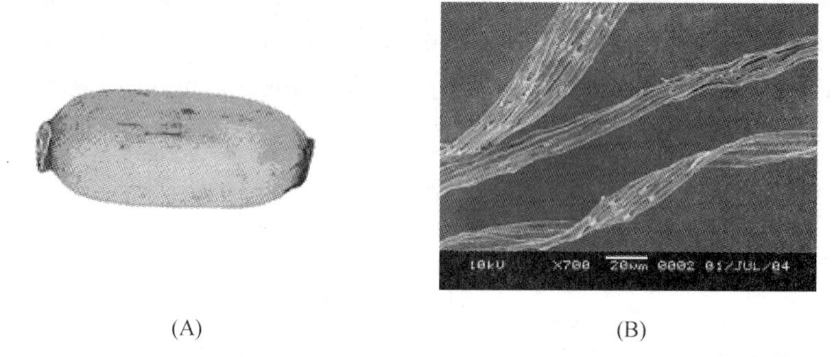

(A) (B)

Figure 3-2 Lotus root fiber: (A) a lotus root, (B) SEM image of the fiber directly hand-drawn from a lotus root

3.13 Medical fiber

The medical fibers are usually those currently hospital applied fibers, e. g. surgery applied fibers. Probably, the nonwoven is also belonged to this area since it is broadly applied as one-time hospital products.

3.14 Biofiber

These fibers are mainly defined for those directly or indirectly made from natural organic raw materials.

3.15 Nanofiber

The nanofiber is only one who was defined based on its diameter size.

3.16 Optical fiber

The optical fiber transmits light where the refractive index varies in the radial direction. Fiber itself is one-dimensional, but light cannot be transmitted unless the structure is two-dimensionally controlled. Optical fiber is a powerful tool to transfer a large amount of information quickly, and plays a key role in supporting today's information technology society. A fine optical fiber like a hair can transmit information equivalent to 6 000 telephone circuits. Although the cost of optical fiber is higher than copper wire, the optical fiber is lighter in weight, higher in capacity and lower in the transmittance loss. An optical fiber is a fine filament, 0.1 mm in diameter, and transmits 95% of input light as far as 1 km. An optical fiber has a two-layer structure of core and clad. A core part is composed of the material with a high refractive index, and a clad part with a low refractive index.

Recommending reading

[1] M. Harris, ed.. *Handbook of Textile Fibers*. Harris Research Laboratories, Inc., Washington, D. C., 1954.

[2] T. Hongu, G. O. Phillips, M. Takigami. *New Millennium Fibers*. CRC Press, Woodhead Publishing Limited, 2005.

Problems

1. On the basis of this chapter described various fibers, how to evaluate the hair of human?
2. If the human's hair can be defined as fiber, which kind it is to be ascribed?

Chapter 4

Typical Fibers and Related Resources

4.1 Fibers from natural polymers

The most common natural polymeric fiber is viscose, which is made from cellulose obtained mostly from plants, e. g. cotton and trees. Other cellulose-based fibers are cupro, acetate and triacetate, lyocell and modal. Few fibers are made from rubber, alginic acid and regenerated protein.

4.1.1 Viscose

There are several fibers made from the naturally occurring polymer cellulose which is present in all plants. Mostly cellulose from wood is used to produce the fibers but sometimes cellulose from short cotton fibers, called linters, is the source. By far the most common cellulosic fiber is viscose.

Viscose is defined very simply by BISFA as being "*a cellulose fiber obtained by the viscose process*". It is known as rayon in the USA. Although several cellulosic fibers had been made experimentally during the 19th century, it became the most popular cellulosic fiber, viscose, was until 1905.

Viscose fibers are made from cellulose from wood pulp. The cellulose is ground up and reacted with caustic soda. After a waiting period, the ripening process during which de-polymerization occurs, carbon disulphide is added. This forms a yellow crumb known as cellulose xanthate, which is easily dissolved in more caustic soda to give a viscous yellow solution. This solution is pumped through a spinneret, which may contain thousands of holes, into a dilute sulphuric acid bath where the cellulose is regenerated as fine filaments as the xanthate decomposes.

Viscose fibers liked the cotton have a high moisture regain. It dyes easily, it does not shrink when heated, and it is biodegradable. It is used in most apparel end-uses, often blended with other fibers, and in hygienic disposables where its high absorbency gives advantages. In filament yarn form it is excellent for linings. It is used very little in home furnishing fabrics but in the industrial field, because of its thermal stability, a high modulus version is still the main product used in Europe to reinforce high speed types.

The structure of viscose fiber is the cellulose structure as below.

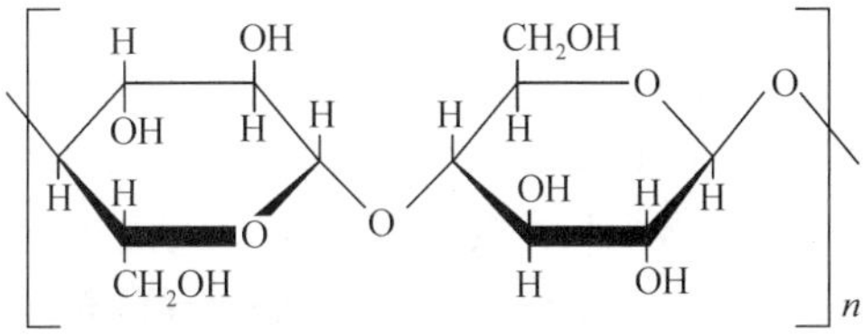

The process for producing viscose fiber was described in below (Figure 4-1).

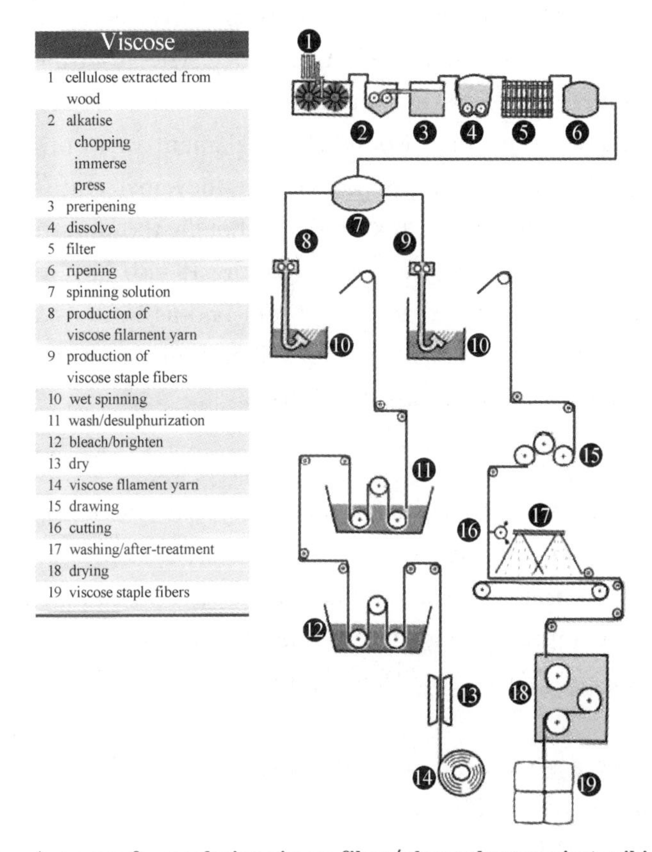

Figure 4-1 A process for producing viscose fiber (chempolymerproject. wikispaces. com)

4. 1. 2 Modal

Modal fibers and polynosic fibers are both high wet modulus fibers made by the viscose process but with a higher degree of polymerisation and modified precipitating baths. This leads to fibers with improved properties such as better wear, higher dry and wet strengths and better dimensional stability.

The cellulose structure is described in below.

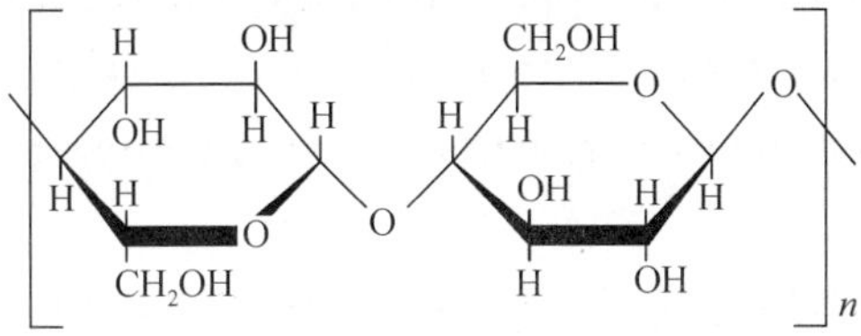

4.1.3 Acetate

The term acetate fiber is used to describe fibers made from cellulose acetate. The difference between diacetate and triacetate fibers lies in the number of the cellulose hydroxyl groups that are acetylated. For acetate fibers the number lies between 75% and 92%, for triacetate fibers it is more than 92%.

Wood cellulose is swollen by acetic acid and then converted to cellulose acetate using acetic anhydride and it is then dissolved in acetone. The resulting viscous solution is pumped through spinnerets into warm air to form filaments. The acetone evaporates and is recovered. The filaments are then wound up as filament yarns or collected as a tow.

These fibers are different from viscose in that they melt, are dyed using disperse dyes, absorb little water and can be textured. Although the dry strength of these two types is similar, triacetate has a higher wet strength. It also has a high melting point, 300 ℃, compared with 250 ℃ for diacetate. Main end-uses for the filament yarns are linings and dresswear. There is very little staple fiber made from these fibers but acetate tow is the major product used for cigarette filters.

4.1.4 Cupro

Cupro cellulosic fiber was first produced commercially in Germany in 1908.

Cotton cellulose is first bleached by boiling in an alkaline solution. This is then dissolved in a mixture of copper oxide and ammonia (the cuprammonium solvent). The blue viscous liquid is pumped through the spinneret into a spinning tube in which weak alkaline water is flowing. This water flow stretches the filaments before they are dried and wound up.

Cupro fibers have a good drape and are easy to wash. The main production is in filament yarn form for woven fabrics, largely for linings.

Cellulose structure is described in below.

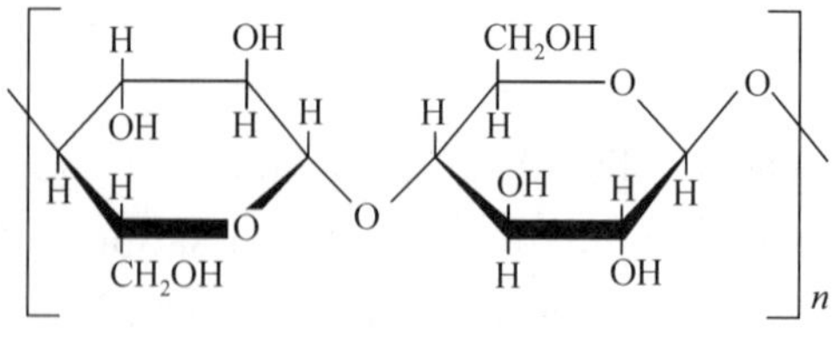

4.1.5 Lyocell

A new generation of cellulosic appeared in the market in December 1992 when a commercial plant in the USA started to make a lyocell staple fiber, based largely on European man-made fiber industry research. Subsequently, two European production plants have opened.

The process used to make lyocell fibers is a solvent spinning process. The cellulose is dissolved in the solvent N-methylmorpholine n-oxide, NMMO, containing just the right amount of water. The solution is then filtered and spun through spinnerets to make the filaments, which are spun into water. The NMMO solvent is recovered from this aqueous solution and reused.

The lyocell fibers, like other cellulosics, are moisture absorbent and biodegradable. They have a dry strength higher than other cellulosics and approaching that of polyester. They also retain 85% of their strength when wet. Under certain conditions lyocell fibers fibrillate which enables fabrics to be developed with interesting aesthetics. Non-fibrillating versions are also available. Lyocell fibers are mostly used for apparel fabrics, especially outerwear, but it has been shown that, due to the fibrillating property some very interesting nonwoven fabrics can be made. Lyocell has the same cellulose structure as above.

4.2 Fibers from synthetic polymers

There are very many synthetic polymeric fibers, e. g. organic fibers based on petrochemicals. The most common are polyester, polyamide, often called nylon, acrylic and modacrylic, polypropylene, the segmented polyurethanes which are elastic fibers usually called as elastanes or spandex and speciality fibers such as the high performance aramids.

4.2.1 Acrylic

BISFA defines acrylic fibers as "*fibers composed of linear macromolecules having in the chain at least 85% (by mass) of acrylonitrile repeating units*". Modacrylic fibers have, in the chain, at least 50% and less than 85% by mass of acrylonitrile. The first commercial fibers were introduced in the USA and Germany in 1948.

The starting materials for acrylonitrile are propylene and ammonia, which are reacted with oxygen in the presence of catalysts. The acrylonitrile is then polymerized to produce polyacrylonitrile, PAN. The PAN is then spun into fibers from a solution in a solvent. Two process routes are used, wet spinning in which the fibers are spun into an aqueous coagulation bath and dry spinning in which the fibers are spun into hot air.

The fibers are then stretched, washed and crimped. The modacrylic fibers contain halogen comonomers such as the vinyl chloride or vinylidene chloride, and have flame-retardant properties.

Acrylic fibers are soft, flexible and have a high loft. For this reason they are widely used in knitted apparel end-uses such as sweaters and socks. In addition to knitted apparel, home furnishing and blankets are other important applications. Acrylic fibers are used as a precursor for producing carbon fiber.

Polyacrylonitrile：

$$\left[CH_2 - \underset{CN}{\overset{H}{C}} \right]_n$$

and acrylic copolymers：

$$\left[\left(CH_2 - \underset{CN}{\overset{H}{C}} \right)_m \left(CH_3 - \underset{Y}{\overset{X}{C}} \right)_n \right]_p$$

The process for fabrication of PAN fiber was described in below by Figure 4-2.

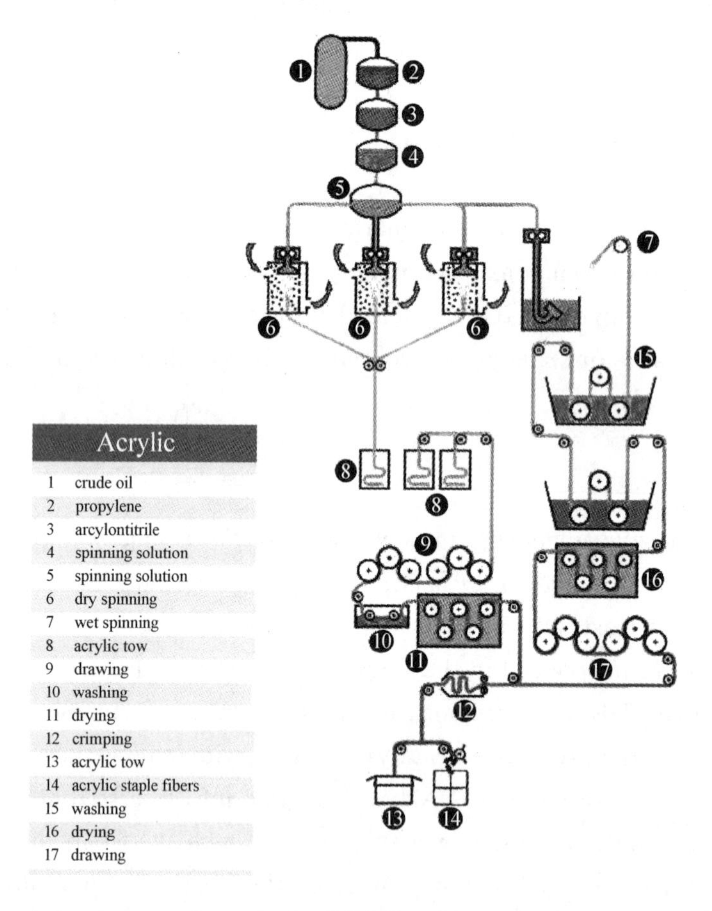

Figure 4-2　A process for producing PAN fiber (chempolymerproject. wikispaces. com)

4. 2. 2　Polyamide

A polyamide fiber is defined by BISFA as being "*a fiber composed of linear macromolecules having in the chain recurring amide linkages, at least 85% of which are joined to aliphatic or cycloaliphatic units*". There are many polyamide fibers made but only two, described below, are made in significant quantities. The first fibers made from polyamide polymers were produced in 1938 in the USA and Germany. In the USA the raw materials, which were used to produce the polymer, were adipic acid and hexamethylene diamine. Since both chemicals contain 6 carbon atoms the new polymer was named polyamide 66. In Germany caprolactam was polymerised to produce a different fiber known as polyamide 6.

Polyhexamethylene adipamide (polyamide 66):

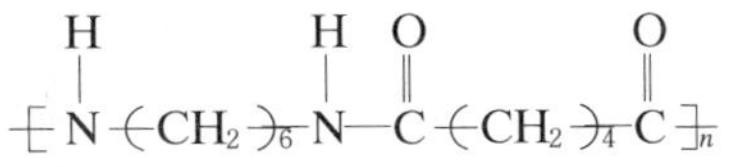

$$\left[\overset{\displaystyle H}{\overset{|}{N}} - (CH_2)_6 - \overset{\displaystyle H}{\overset{|}{N}} - \overset{\displaystyle O}{\overset{\|}{C}} - (CH_2)_4 - \overset{\displaystyle O}{\overset{\|}{C}} \right]_n$$

Polycaproamide (polyamide 6):

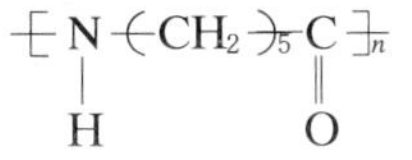

$$\left[\underset{\displaystyle H}{\underset{|}{N}} - (CH_2)_5 - \underset{\displaystyle O}{\underset{\|}{C}} \right]_n$$

To produce fibers from the polyamide polymers the molten polymer is pumped through spinneret holes at a temperature approaching 300 ℃ to form filaments that are cooled and solidified in a quench air stream. If filament yarn is being produced the filaments are then oiled and wound onto cylinders. Polyamide yarns are spun to different orientations depending upon the use. If a fully oriented yarn (FOY or FDY) is required it is achieved by having a draw stage on the spinning machine, a process called spindrawing, or by spinning the yarn at very high speeds.

If the yarn is to be textured the preferred orientation is partial, POY. This yarn is then fully drawn and textured in a separate process. If staple fibers are being produced, very many filaments are bundled together to form a tow which is subsequently stretched, crimped and cut to the desired length. In 1999 there were over 3.4 million tons of polyamide filament produced worldwide and over 0.5 million tons of polyamide staple.

In weaving the main end-use is for outerwear and technical fabrics. In knitting, stockings and tights and outerwear are both important outlets. Carpets and ropes and twines are also important sectors.

A process on preparation of the PA fiber was presented in Figure 4-3.

4.2.3 Polyester

A polyester, according to BISFA, is "*a fiber composed of linear macromolecules having a chain at least 85% by mass of a diol and terephthalic acid*". The first polyester was made in the UK in 1941. This polyester, known as polyethylene terephthalate, PET, has become the world's major man-made fiber. Other polyesters such as polybutylene terephthalate, PBT, and polytrimethylene terephthalate, PTT, are made but in much smaller quantities. The structure of these three polyesters are presented in below.

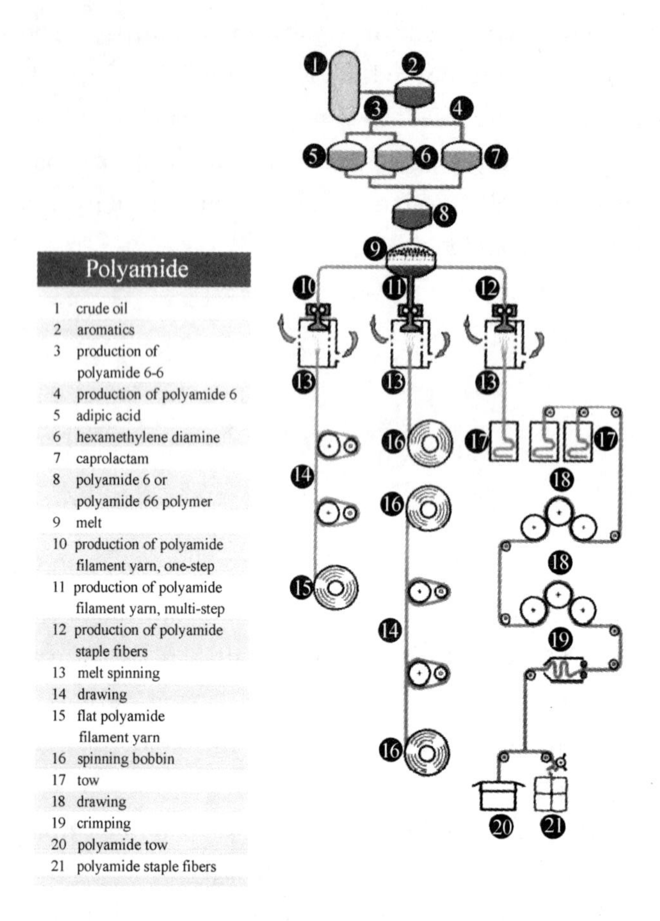

Figure 4-3　A process for producing PA fiber (chempolymerproject. wikispaces. com)

$$-\!\!\left[O - \overset{\overset{O}{\|}}{C} - C_6H_4 - \overset{\overset{O}{\|}}{C} - O - CH_2CH_2 \right]_n \qquad \text{PET}$$

$$-\!\!\left[O - \overset{\overset{O}{\|}}{C} - C_6H_4 - \overset{\overset{O}{\|}}{C} - O - CH_2CH_2CH_2 \right]_n \qquad \text{PTT}$$

$$-\!\!\left[O - \overset{\overset{O}{\|}}{C} - C_6H_4 - \overset{\overset{O}{\|}}{C} - O - CH_2CH_2CH_2CH_2 \right]_n \qquad \text{PBT}$$

Polyester fibers are made in a very similar way to polyamide. Some plants take polyester polymer chips and melt them, at around 280 ℃ and then extrude the melt into continuous filaments to be wound onto packages or collected in cans as a tow before being stretched, crimped and cut into staple fiber. Other plants produce the polymer by a continuous process, CP, and form it into fibers without producing chips. A growing quantity is made by recycling PET bottles and other waste. If fully oriented yarns, FOY, are being produced the fibers are drawn on the spinning machine. If the yarn is to be textured, partially oriented yarns, POY, are spun.

In Western Europe, apparel accounts for a large share of usage of polyester fibers. Industrial use, such as tire fabrics, and unspun uses such as furniture fillings and nonwovens, are both expanding rapidly.

Poly (ethylene terephthalate):

$$\left[\overset{O}{\overset{\|}{C}} - C_6H_4 - \overset{O}{\overset{\|}{C}} - O - CH_2 - CH_2 - O \right]_n$$

The PET fiber was spun and prepared according to below scheme (Figure 4-4).

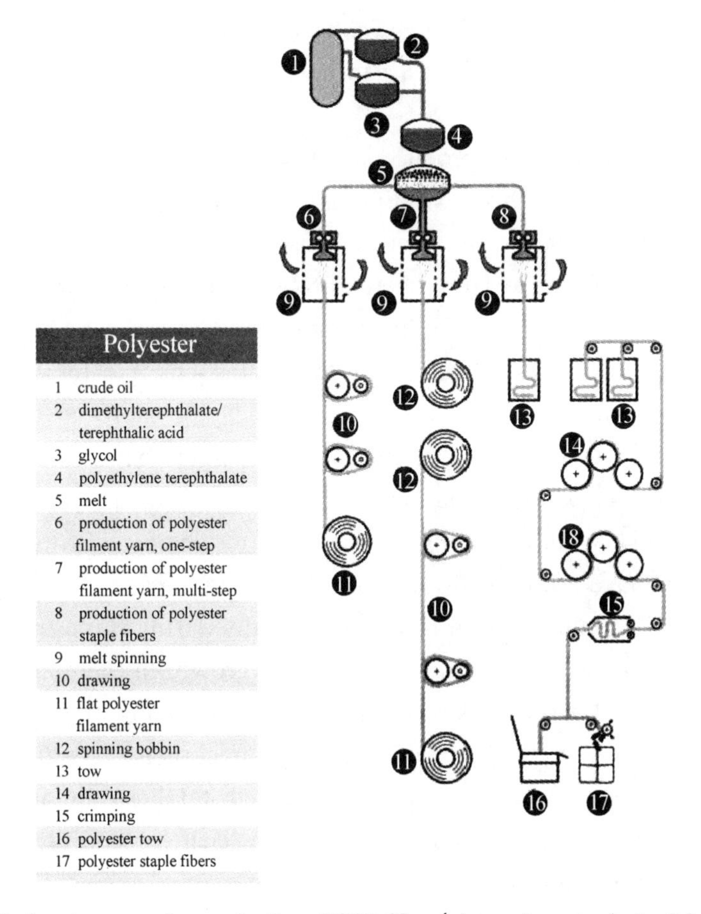

Figure 4-4 A process for production of PET fiber (chempolymerproject. wikispaces. com)

4.2.4 Polyolefins

There are two polyolefin polymers used to make synthetic fibers, polypropylene and polyethylene, with polypropylene being by far the most important. The BISFA definition for polyethylene fibers is "*fiber composed of linear macromolecules of unsubstituted saturated aliphatic hydrocarbons*" and for polypropylene fibers "*fiber composed of linear macromolecules made up of saturated aliphatic carbon units in which one carbon atom in two carries a methyl side group...*". Polyethylene was first produced in the UK in 1933 by polymerizing ethylene under pressure. In 1938 in Germany polyethylene was made by polymerizing ethylene in an emulsion. Polypropylene was commercialised in 1956 by polymerizing propylene using catalysts. Both of these polyolefins are very important in plastic moulding and for making plastic sheet but both are spun into synthetic fibers on a large scale.

Polyolefin fibers are made by melt spinning. Usually polymer granules-made by specialist producers rather than fiber companies-are fed to an extruder which melts the polymer which is then pumped through a spinneret. The filaments are cooled in an air stream before being wound on a package or collected in cans as a tow. Because the fibers are difficult to dye, colored pigments are often added to the polymer stream before extrusion.

An alternative process is to produce a film, cut the film into strips and then fibrillate the individual strips before winding onto a package. Recently a new family of catalysts to make polypropylene has been developed called metallocene catalysts. It is claimed that the polymers made from these catalysts can be spun to finer counts and drawn to give higher tenacities than existing polymers.

Both polyolefin fibers have a density less than 1.0 g/cm^3 and therefore, at a given decitex, are thicker than other man-made fibers and give more cover. They do not absorb moisture, which is an advantage in many end users, but without modification, they cannot be dyed. Their melting points are around 130 ℃ for polyethylene and 160 ℃ for polypropylene. They have a high resistance to chemical attack and modern polypropylene fibers have a high resistance to UV degradation.

Polypropylene fiber consumption has grown rapidly during the past decade. This is due largely to its acceptance as a carpet fiber and the growth in the nonwoven end-uses, especially disposables and geotextiles where polypropylene is now the dominant fiber.

In 1998, 370 000 tons of polypropylene were used in carpets, 228 000 tons in hygiene and medical and 106 000 tons in geotextiles and agrotextiles. In addition to these products the properties of the polyolefins make them ideal for end uses such as ropes, tapes, twines, fishing nets and sacking (from slit film).

Polypropylene:

$$\left[CH_2 - \underset{CH_3}{\overset{H}{\underset{|}{\overset{|}{C}}}} \right]_n$$

4.2.5 Aramid

A special fiber can be described as being a fiber with unique properties which make it the preferred fiber for particular applications. They are more expensive than the other synthetic fibers described in this document and are produced in comparatively small volumes. Statistics about the production and consumption of these fibers are not widely available.

There are very many specialties, or niche, fibers.

Aramid is a contraction of aromatic and polyamide. BISFA defines these fibers as "*fiber composed of linear macromolecules made up of aromatic groups joined by amide or imide linkages*". There are two types of aramid: the meta-aramids and the paraaramids. The development of aramid fibers took place during the mid-1950's to mid-1960's in the USA. The fibers were shown to have high melting points and high moduli. The impetus to develop the fibers came for the need for high performance fibers for air and space travel.

The polymer polymphenyleneisophthalamide is used to make meta-aramids and the polymer polypterephthalamide to make paraaramids. Because the aramids decompose before they melt they are produced by wet and dry spinning methods. Sulphuric acid is the normal solvent used in the spinning processes. In wet spinning, a strong solution of the polymer, which also contains inorganic salts, is spun through a spinneret into weak acid or water. In this bath the salts leach out. In the dry spinning process the salts are more difficult to remove and this process is only used to produce the weaker meta-aramid fibers. In both processes post treatment of the fibers by additional drawing is used to optimize fiber properties. Aramid products are available as filament yarn, staple fiber or pulp.

Some of the main end-uses for meta-aramids are protective clothing, hot gas filtration and electrical insulation. Para-aramids are used to replace asbestos in brake and clutch linings, as tire reinforcement, and in composites such as materials for aircraft, boats, high-performance cars and sports equipment. Members of police forces and armed forces wear anti-ballistic aramid apparel.

4.2.6 Elastane

Elastane yarns are characterized by their ability to recover from stretch. BISFA describes them as "*a fiber composed of at least 85% by mass of segmented polyurethane which, if stretched to three times its unstretched length, rapidly reverts substantially to the unstretched length when the tension is removed*". Although elastane was first synthesized in 1937, it was not commercialized as a fiber until 1958.

In addition to their remarkable stretch and recovery properties, elastanes resist perspiration and cosmetic oils, are easily washable, are dyeable and have moderate abrasion resistance. Elastane yarns are often covered with another fiber before use. This provides more bulk and improves abrasion resistance. The main end-uses for the yarns are garments and other products, where comfort and/or fit are important.

4.3 Fibers from inorganic materials

The inorganic man-made fibers are fibers made from materials such as glass, metal, carbon or ceramic. These fibers are very often used to reinforce plastics to form composites.

4.3.1 Glass

There are many inorganic fibers, including glass, carbon, metal and ceramic. They are used particularly in the industrial fiber sector. Two important inorganic fibers are described in this section. Glass is the most important inorganic fiber.

It is produced by melting glass pellets in an electric furnace at around 1 500 ℃. The molten glass passes through small holes in a plate at the base of the furnace. After cooling in air it is wound up on a package. It can be alternatively spun centrifugally to form web type.

There are several types of glass fibers have been produced. They have high moduli, high rot resistance, low moisture uptake and are brittle with low breaking extensions. Glass is used extensively for insulation cases in the form of a felt and has been applied for reinforcing plastics to make boats, caravans, automobile parts, flame-resistant curtains and decor fabrics.

4.3.2 Carbon

The BISFA definition of a carbon fiber is "*fiber containing at least 90% by mass of carbon obtained by thermal carbonization of organic fiber precursors*".

The common precursors used to make carbon fibers are polyacrylonitrile, PAN, pitch, viscose and lignin. If pitch is used the process consists of extrusion, oxidation and graphitization. If PAN is used, a tow is oxidized, then carbonized followed by graphitization. On lignin-based carbon fiber, a selection is required because lignin is a byproduct of pulping process which could be performed using various chemical methods.

Carbon fibers are characterized by having high moduli and high strength, especially when embedded in a matrix such as epoxy resin. They are also brittle and have a low density. The main end-users are as reinforcement fibers in composites for the aircraft and aerospace industries and sports goods.

In general, the carbon fiber is also involving the activated carbon fiber, ACF.

Recommending reading

[1] M. Harris, ed.. *Handbook of Textile Fibers*. Harris Research Laboratories, Inc., Washington, D. C., 1954.

[2] J. G. Cook. *Handbook of Textile Fibers*, 5th ed.. Merrow Publishing, Durham, N.C., 1984.
[3] T. Hongu, G. O. Phillips, M. Takigami. *New Millennium Fibers*. CRC Press, Woodhead Publishing Limited, 2005.

Problems

1. How to ascribe the resource of a human hair?
2. Please ascribe the resource of a lignin fiber.

Chapter 5

Main Fiber Formation Techniques

Fiber directly from synthesis process is usually regarded as the one step process. Otherwise, there are at least two steps, because the synthesized polymers required spinning to form fiber. In this section, we described the method of fiber spinning by two parts, the first is focusing on the general equipments for polymerization, and the second is on various spinning techniques.

5.1 General equipments and components for synthesis of polymers

5.1.1 Autoclaves

These reactor vessels usually consist of a cylindrical vertical vessel with a flanged dished cover, lower dished bottom or conical bottom, jacket and/or heating coils for heating or cooling and the agitator. Due to the large variety of forms only the combination of the possibilities will be mentioned. Examples for the most important applications are given. Heating jackets are recommended for vapor with condensation while jackets with spiral or welded half pipes are mostly for liquid heaters.

The type of autoclaves can be known from Figure 5-1 and those can be defined as two

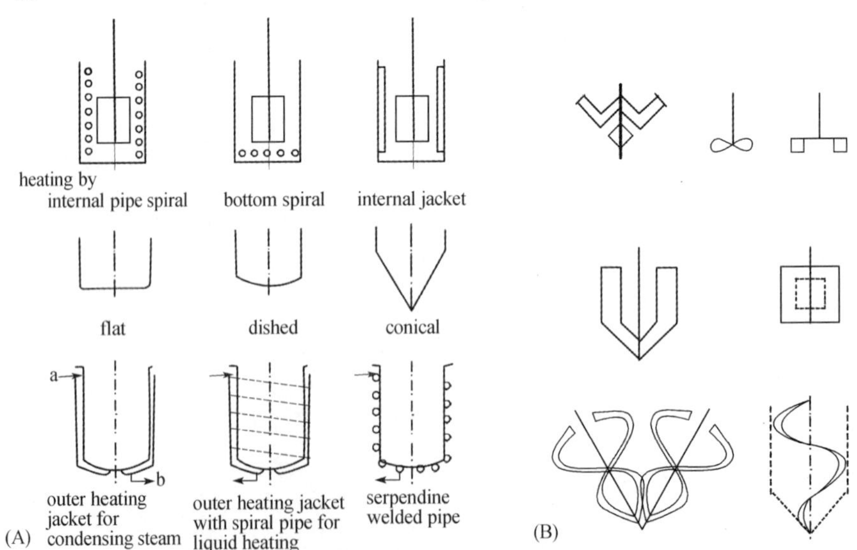

Figure 5-1 Typical autoclave shapes: heating by liquid or vapor (A), and agitator types (B) (F. Fourne. *Synthetic Fibers, Machines and Equipment, Manufactory, Properties.* Hanser Publisher, Munich, Germany, 1998)

series, i.e. one is the continuous type and another is the batch type (Figure 5-2, left).

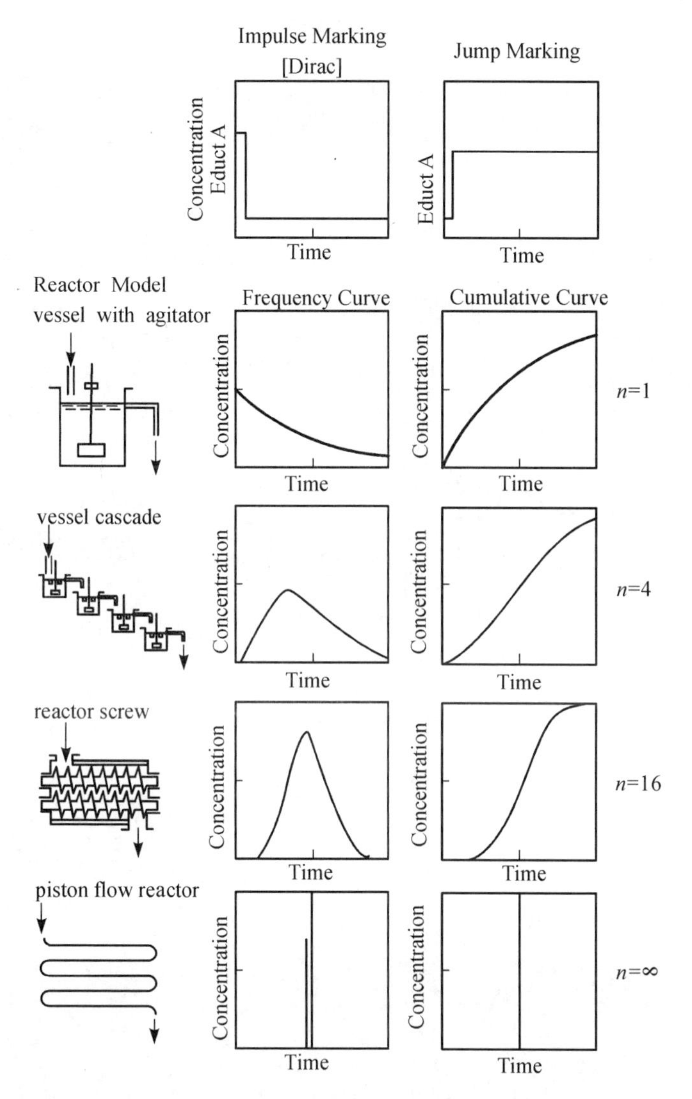

Figure 5-2 Dwell time distribution for different reactor models and comparison with a reactor screw (F. Fourne. *Synthetic Fibers*, *Machines and Equipment*, *Manufactory*, *Properties*. Hanser Publisher, Munich, Germany, 1998)

5.1.2 Related mechanical components

As above mentioned that a whole autoclave system posses several components, e.g. a vessel with some holes to fit the viewing, measuring and adding additives, an agitator, a heating system, and something to be required for detailed operation.

5.2 Melt spinning process related machines and equipments

The melt spinning is the most popular method broadly applied in factory for producing polymer fibers. The process could be briefly described as below by three photographs (Figure 5-3) and a scheme (Figure 5-4).

Figure 5-3 Three floor polyester filament spinning plant: (A) spinning extruder deck with insulated melt distribution piping; (B) spinning heads with air quench chambers; (C) ground floor with POY take-up machines (with godets)

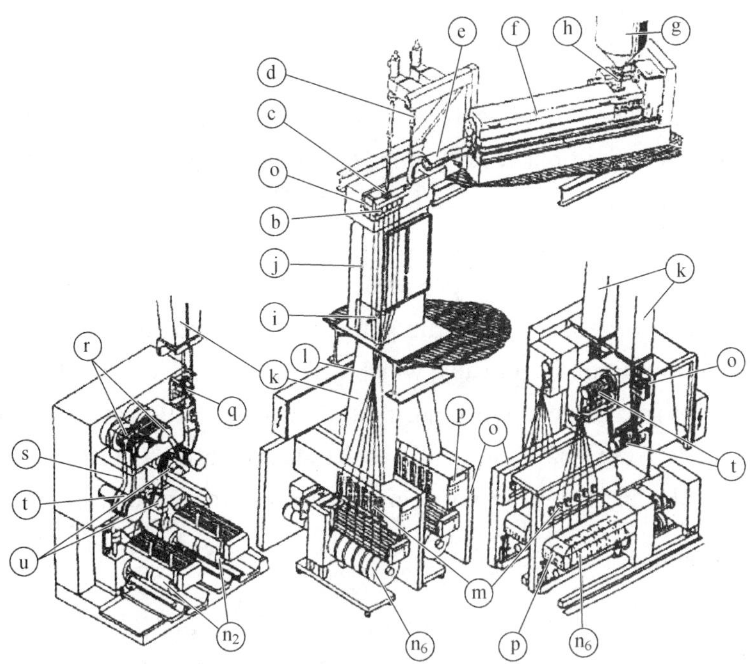

(F. Fourne. *Synthetic Fibers, Machines and Equipment, Manufactory, Properties*. Hanser Publisher, Munich, Germany, 1998)

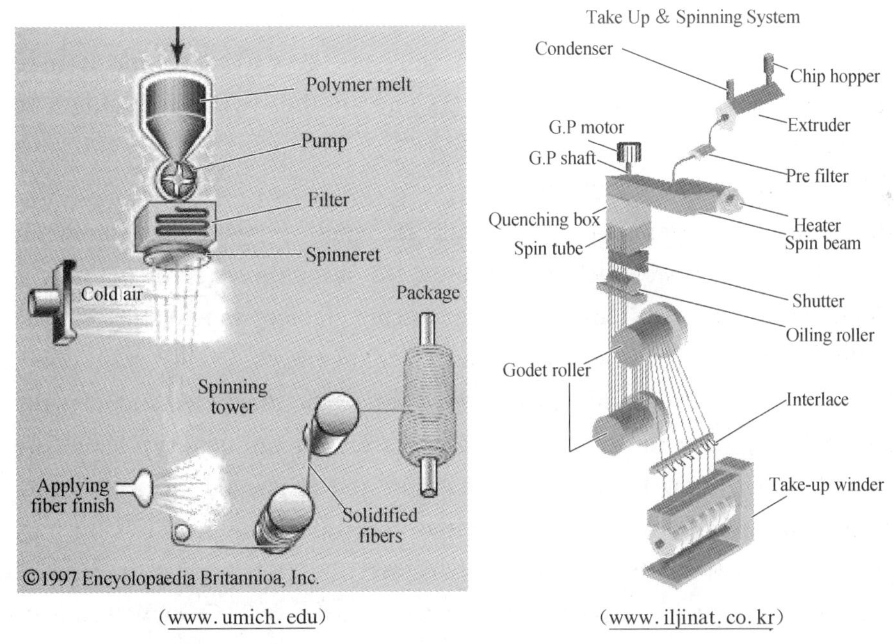

(www.umich.edu) (www.iljinat.co.kr)

Figure 5-4 Principle design of a melt spinning machine (Barmag AG): a) spinning beam; b) spin packs; c) spinning pumps; d) spinning pump drives; e) non-stop filter system; f) spin extruder; g) chip hopper; h) chip gate valve; i) air quench chamber; j) spin finish application in quench; k) floor interconnection tube; l) turning of the filament plane; m) yarn sensor system; n) high speed winders for textile; o) noise absorbent walls; p) winder control elements; q) spin finish system on the take-up machine; r) godets, hot or cold, with idler rolls or godet duo; s) BCF texturing aggregate; t) BCF cooling drum; u) filament pretensioning system; v) melt supply pipe; x) drive for quendi-applied spin finish; y) conditioned supply air; z) dowtherm (Diphyl) loss condenser; n_2) bobbin take-off and transport; n_6) revolver winder for 6 packages

5.2.1 Chip gate valves

There have two types for chip gate vales, one is the manually operated with the tank connection above, two solid plates, a frame for the gate positioned on the inside and the spindle sealed pressure and vacuum proof towards the outside. In the transport direction it is chip proof, but not pressure and vacuum proof. Vacuum proof in the direction of the chip transport as well as to the outside up to 10～3 mbar is the chip gate valve. Turning of the gate lever would cause the gate opening or closing.

The second is a traditional ball valve which is frequently used while less recommended, because they are not free of dead space during the turning of the ball, and the seals to the ball are usually made from PTFE or similar and thus are softer than the chips. Therefore they can easily be damaged and do not remain pressure and vacuum proof.

5.2.2 Spin extruders

Spin extruder is a main machine within the melt spinning process and it usually has several types, e.g. the single screw-based or multiscrews-based. In general, the spin extruder can be divided into several areas such as the feeding, metering, compressing, forced transporting and mixing.

Two parameters are usually applied to describe the extruder, one is the ratio of length, L, to diameter, D, defined as L/D, and another is the area length for above mentioned zones.

5.2.2.1 Single-screw spin extruder

Asides from the aforementioned details the melt pressure and temperature are measured as closely as possible behind the screw tip in the direction of the flow as can be seen in Figure 5-5. The melt temperature measuring element should stick at least 10～20 mm into the melt to avoid measuring undefined averages to the wall temperature. Nitrated materials for screws and cylinders as usual for plastic extruders will corrode after some operating time and significantly lower the filament quality. Therefore usually bimetal cylinders with corrosion resistant coating and if needed screws from corrosion resistant materials are used which are ionitrated. The operating pressures for most textile raw materials are between 80 bar and 150 bar, and the melt temperatures are between 220 and 320 ℃. In special cases up to 500 bar and/or 500 ℃ may be needed. The operating rpm for extruders of about 50 mm screw diameter are adjustable between 20 and 130 r/min and for $D = 250$ mm between 15 r/min and 80 r/min. Considerably larger single-screw extruders were not successfully introduced for spinning, because the necessary melt distribution ways are too long as are the resulting melt dwell times.

5.2.2.2 Double-screw extruders

Only very few companies use it as a melt extruder for spinning filaments, and it is usually replaced in new spinning plants by single screw spin extruders. Nowadays the double

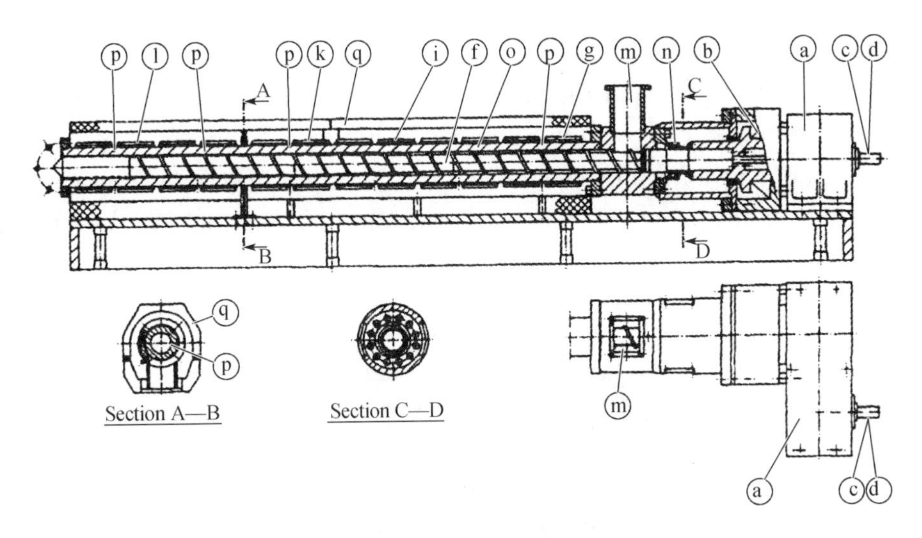

Figure 5-5 **Spin extruder: a) reduction gearbox; b) thrust bearing; c) V-belt drive; d) slipping and/or overload clutch in; f) extruder screw (cylinder); g, l) heaters and heating zones; m) feeding zone; n) shaft packing (sealing); o) extrude barrel; p) temperature measuring elements; q) extruder barrel; q) insulation cover (F. Fourne. *Synthetic Fibers, Machines and Equipment, Manufactory, Properties.* Hanser Publisher, Munich, Germany, 1998)**

screw extruder is primarily used to transform powder to chips, to compound, i. e., to enter additives and/or dyes into the melt, e. g., to produce master batches, and to mix several melted polymers. It is also used as a reactor for plastics preparation and as a finisher in the final step of a poly condensation, usually with an increased volume. For melt spinning the normal volume double screw extruder with synchronized combing screws is almost used exclusively with the following advantages: very good self cleaning, distribution and division effects, pressure guidance, even under high pressure or vacuum, and good lengthwise and cross-sectional mixing effects. The double jacket housings can easily be heated. The screws allow a high driving force to be entered into the polymer. However, they can only be used for melt dwell times under 7 min and due to their construction in comparison to a single screw extruder of similar power they are relatively expensive.

The effects of the double screws: self cleaning of the screws by reciprocal stripping motion, transport in a system that is lengthwise open: The screw threads form continuous channels from the entry to the tip. Transport characteristic mostly represents a drag flow with only partial forced flow, resulting in a very narrow dwell time distribution yet good mixing of the melt.

5.2.3 Spinning heads and spinning beams

Spinning heads are loaded with granulate, melt it and dose its distribution to the spinnerets. Spinning beams are loaded with melt and dose its distribution also to the spinnerets. Production spinning heads are dow vapor heated to about 320~340 ℃, as are spinning beams.

The fork like distribution of the melt flows results in the shortest dwell times, but in unequal dwell times if one of the spinning positions is stopped. The star shaped distribution has the advantage of equal dwell time-even when deactivating one spinning position-but the disadvantage of the longest dwell time on the way to all spinning positions. For that reason often a mixed pipe layout is chosen (Figure 5-6).

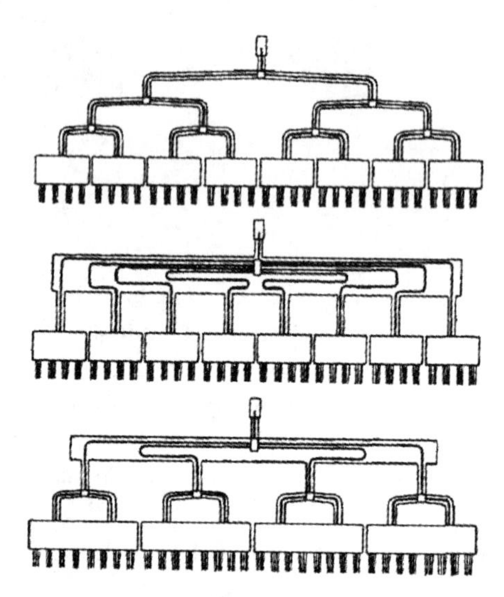

Figure 5-6 Possible melt manifold configurations for spinning beams and/or to spinning positions. Top: forked distribution behind each joint; Middle: central distribution with equal pipe lengths to each spinning position; Bottom: mixed system (F. Fourne. *Synthetic Fibers, Machines and Equipment, Manufactory, Properties.* Hanser Publisher, Munich, Germany, 1998)

5.2.4 Spinning pumps

As spinning pumps (Figure 5-7) only gear pumps of extreme exactness are used that can transport the melt or solution flow etc. with constant volume and pressure and at the same time even out variations in the supply-even at increasing pressure loss of subsequent filters, e.g. in front of the spinneret. Thus per spinneret a separate pump stream is needed because the hydrodynamic separation is not exact enough. Spin pumps are combinations of plates/gears and shafts with manufacturing tolerances of <0.5 μm and therefore have to be handled with care. Smallest dust contamination between the plates already results in pump blocking.

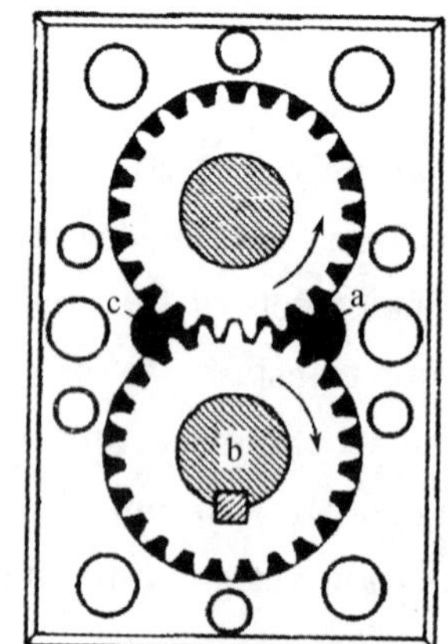

Figure 5-7 Working principle of a gear pump for a melt metering. a) melt entrance; b) gear driveshaft; c) melt discharge

In this process, the important spinning conditions are:

(1) macromolecular chains with a minimum of branches and networks with more than a minimum degree of polymerization;

(2) sufficient ability of the melted mass to be drawn into filaments, as for example defined by the filament breaking length. The latter depends on the viscosity and the draft velocity. The maximum indicates the optimum spinnability of the material. Too high a take-up speed and/or too low a viscosity leading to break filaments (e.g. due to too thin melt or too high spinning temperatures). Too high molecular weights or viscosity or too fast hardening or coagulation of the outer layer lead to a melt fracture, e.g. melting rupture (=saw toothed appearance).

Melt spinning is done for most polymers (PA, PET, PP et similes) between 240 ℃ and 320 ℃, for some special polymers under 200 ℃ or between 350 ℃ and 450 ℃, usually under high pressure (100~200 bar) through very thick spinneret plates (=10 mm).

The broadly used materials for producing spinning pump are described as below, where some kind of materials are selected mainly depending of the polymer melting-based temperature (Table 5-1).

Table 5-1 Recommended materials for spinning pumps

Steel containing	For			
	Polyester, PP	Polyamid	Polyramid	650 ℃, 700 bar
C	2.0	1.25		
Cr	12.0	4.3		
M	0.85	0.9		
W		12.0		
V		3.8		
Ni		—	Hastelloy C 276	CPM 10-V CPM-T-15
	Cr 12 1.320 2	similar to 1.260 1	(Zenith)	(Zenith)

5.2.5 Spin finish pumps

Spin finish pumps correspond in their design to the spin gear pumps with 1 or 2 exits per gear level and up to 12 dosing streams per pump. Because of the low viscosity of the spin finishes, usually water-spin finish-emulsions, and because of their low lubrication effect the manufacturing precision has to be even higher than for spin pumps.

5.2.6 Discharge pumps and "in-line" pumps

The former described of its function and the latter described of its application place. In many cases, the former pump can replace the traditional discharge screws and the melt extruders, while the latter is able or unable depending on the linked component at the middle (unable) or at the final place (able).

5.2.7 Melt and solution filters

Various filter media assembles were found applie in industry such as sand filter above stainless steel wire mesh filter, Fuji filter, sintered metal powder, sintered metal fiber nonwovens. The type can be seen as below (Figure 5-8).

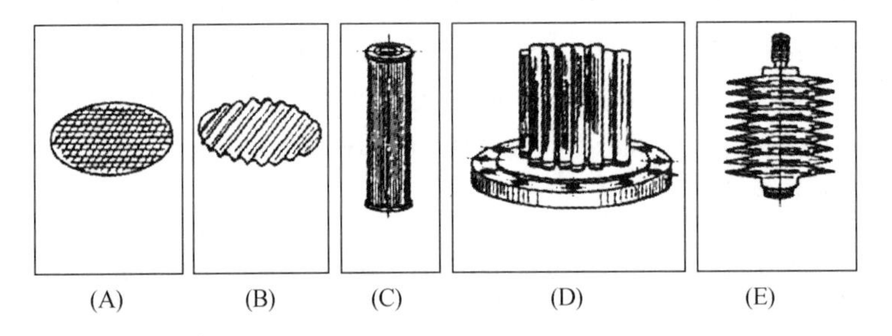

Figure 5-8 Various filter types. (A) flat filter; (B) pleated flat filter-cylinder; (C) pleated cylinder-candle; (D) many small cylinders-candles; (E) biconical disk filter assembly

5.2.8 Spinnerets

Spinneret is the actual filament-forming component and shape controller. It converts the melt or solution into the required number of filaments with designed shape as determined of the filament cross-section (Figures 5-9, 5-10 and 5-11). For high melt inlet pressures-based spinneret, the plate thickness is within 8～25 mm, and plate thickness of 0.8～2 mm is used for low inlet pressures. For solutions and melts the area-based holes at >50 holes/cm^2 is usually recommended and used.

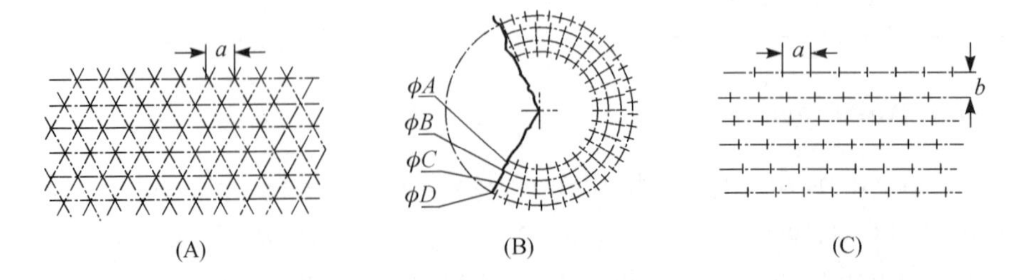

Figure 5-9 Hole layouts for melt spinnerets. (A) densest (staggered) arrangement; (B) triangular pitch arrangement for round spinnerets: radial spacing and tangential spacing; (C) slightly staggered hole rows

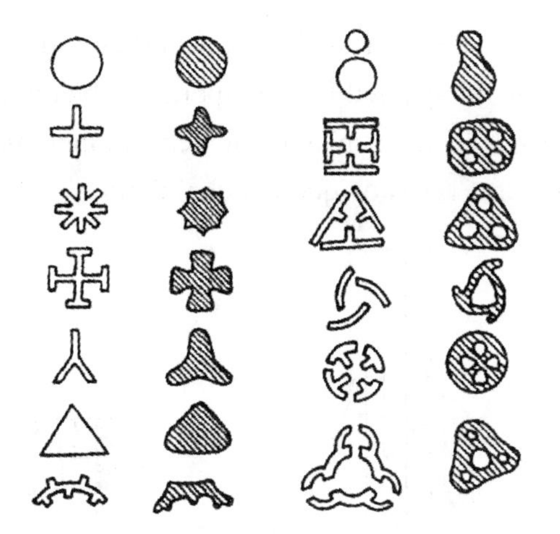

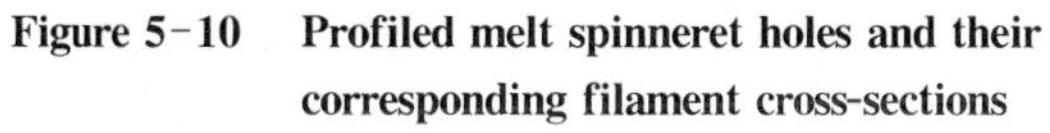

Figure 5-10 Profiled melt spinneret holes and their corresponding filament cross-sections

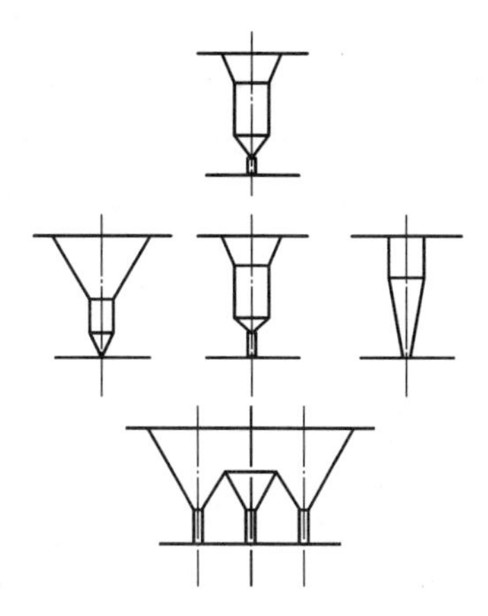

Figure 5-11 Cross-sections of typical melt spinneret capillaries

Table 5-2 presented various materials recommended for producing spinnerets

Table 5-2 Properties of spinneret materials

Material	T(℃)	$\sigma_{0.2}$(kg/mm^2)	Hardness
Au Pt Ir		2~5	≈250 HV
Ni		10~21	
Hastelloy C		>32(<420 ℃)	
Ta			110 HV (treated 300 HV)
14 122	300	54	
14 550	300	14	
1 457	300	15	130~190 HB
14 580	300 (450)	15	
A S 17/4 PH			270~290 RC/=450 HB
Pt-Rh-80/20	700~1 350	8+3	

Damaged spinnerets should definitely be returned to the manufacturer, as the cost of material is 150~200 DM/g (1985/1986's price), while the manufacturing cost is less than 10% of this. Spinnerets made from sintered Al_2O_3 plates have not proved successful due to the edges of the melt exited holes abrade rapidly.

Glass spinnerets have also been used for normal solution wet spinning processes.

5.2.9 Spin packs

Spin packs incorporate, in one housing, the components in the flow direction: pack top cover (including melt distribution), pack filter, support or distributor plate and the spinneret. It must also be possible to exchange the spin pack quickly and easily.

5.2.10 Quench cabinets

Filaments extruded from the spinneret capillaries are quenched as single filaments almost parallel to one another, and are solidified under almost time-constant conditions, without filament flutter or fusion; they are-if possible-cooled to a temperature below the glass transition temperature. For these reasons, the quench and quench cabinet must meet certain criteria. Figure 5-12 presented several types of quench chambers.

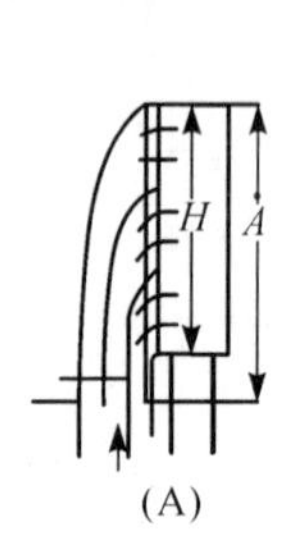

(A)

Turblent chamber
$H\approx 1.4...2.2$m
Zone air velocity profile set by hinged inlet vanes
$Q>20\times G_{melt}$(kg/h)
$v>0.4$ m/s
Step profile

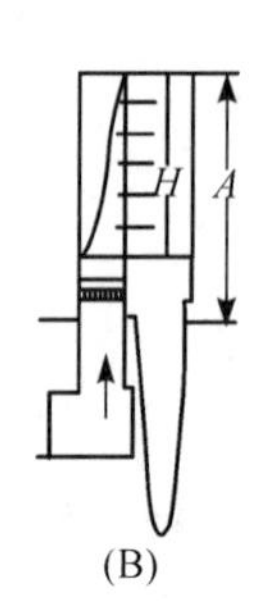

(B)

Standard quench chamber
$H\approx 0.4...0.8$ m
Air velocity profile set by a deflector plate
$Q(\mathrm{Nm^3/h})\approx(15...30)G_{melt}$
$v=0.1...0.5...0.7...0.3$ m/s
Continuous velocity profile

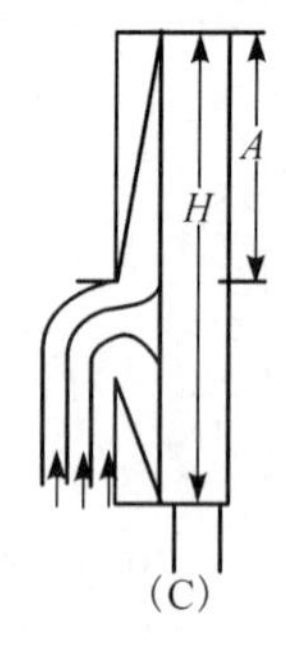

(C)

Extra long quench chamber
$H>2.5...5$ m
Air velocity profile set as in (B),for 2 or 3 zones
Q_2,above$\approx$35 G
Q_{2+}below$\approx$12 G
$v\leqslant 1.3$ m/s
Preferred for POY when final dpf$\geqslant$8. . . 20

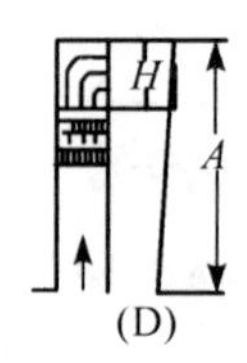

(D)

Turbulent 1-chamber
$H\leqslant 0.4$ m
$W(H)$=constant
$Q(\mathrm{Nm^3/h})\leqslant 40\ G_{melt}$
$v\geqslant 0.8$ m/s
Preferably $v>1.5$ m/s for POY spun at high yarn tension

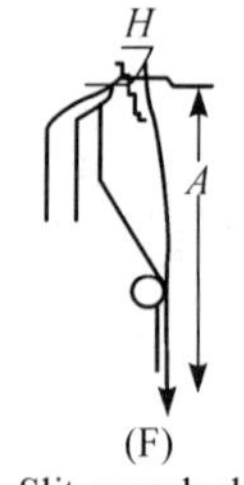

(F)

Slit quench chamber
$H<0.05$ m
$Q\geqslant 40$ G
$v\geqslant 10$ m/s
For PET and PP compact staple fiber spinning

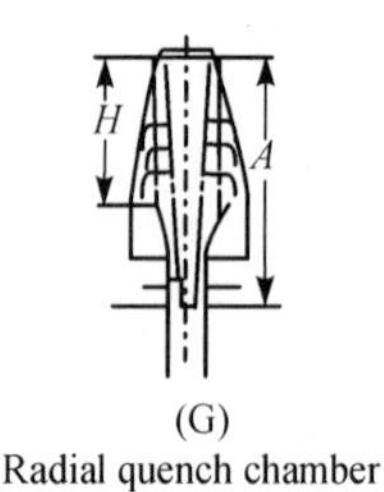

(G)

Radial quench chamber
(inside→outside) $H\leqslant 1.2$ m
$v\approx 0.4...1.0$ m/s
$Q\approx 8$ (for PET)...25 (for PA) $\times\ G$ (kg/h)
Preferred for staple tow (in vertical spinning)
Number of spinneret holes>2 000

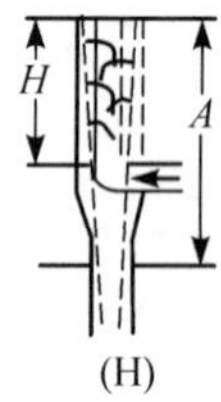

(H)

Radial quench chamber
(outside→inside)
$H\leqslant 1.2$ m
$Q\approx 16$ G
$v<1.0$ m/s
weakly turbulent

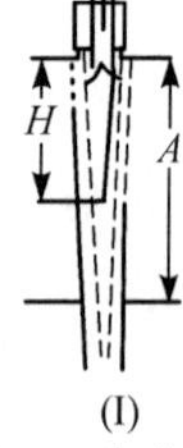

(I)

Radial quench chamber
As H, but with air inlet from above, through core of spinneret

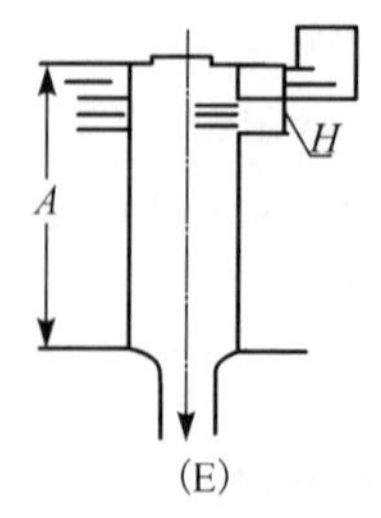

(E)

Highly turbulent quench chamber
$H\approx(2\text{ or }3)\times 0.1$ m
Countercurrent (!)
$Q\approx 36\times G$
$v>6$ m/s
For high spinneret hole density PET staple fiber spinning

Figure 5-12 Various quench chamber types (F. Fourne. *Synthetic Fibers, Machines and Equipment, Manufactory, Properties*. Hanser Publisher, Munich, Germany, 1998)

5.3 Wet spinning

In wet spinning (Figure 5-13), the polymer solution was prepared in a suitable solvent then be extruded as a fiber into a coagulation bath containing a nonsolvent. The limits of polymer concentration in the spinning solvent are determined by polymer solubility and solution spinning pressure limitations. The polymer concentration used for wet spinning is lower than that in dry spinning due to the solution spun at lower temperatures. Of the wet spinning, the spinneret was submerged in the liquid coagulation bath, and the emerging filaments are coagulated in a precipitating bath or a series of baths of increasing precipitant concentration. During the short residence time in the coagulation bath, the fiber formed and its structure taken shape as a result of complex. In this process, the counter-diffusion of solvent and nonsolvent and phase separation of the polymer would be taken place.

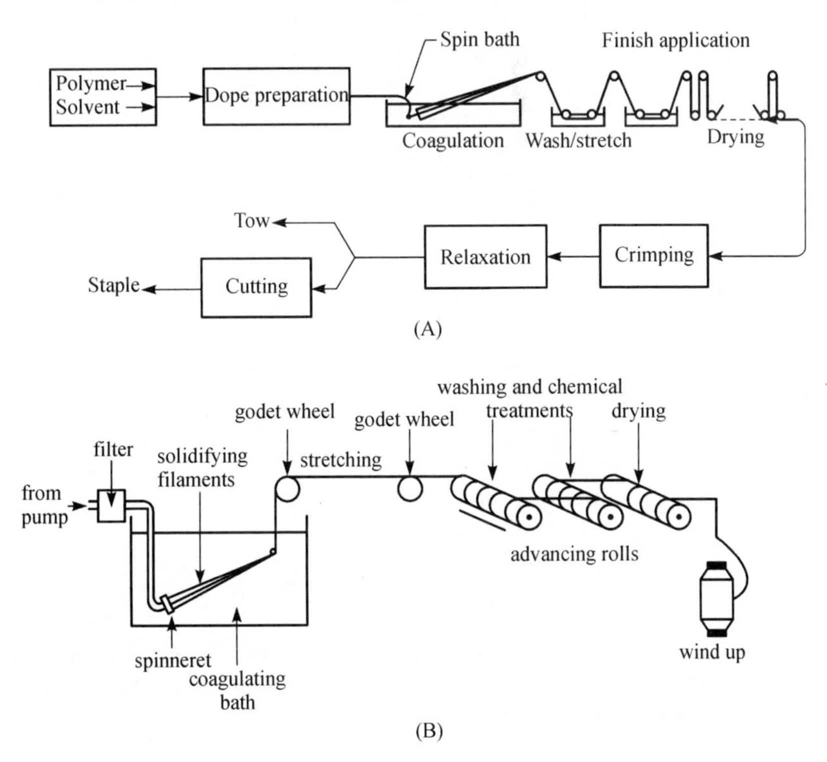

Figure 5-13 A scheme of the wet spinning (vivianbuff. com)

5.4 Solution dry spinning

During the solution dry spinning (Figure 5-14), the polymer spinning solution is initially metered at a constant temperature by a precision gear pump through a spinneret into a cylindrical spinning cell of 3~8 m in length. Heated cell gas, made up of solvent vapor and an inert gas, e.g. nitrogen, is therefore introduced at the top of the cell and passed through a distribution plate behind the spinneret pack. Since the cell gas and cell walls are maintained with high temperature, this leads the solvent evaporation rapidly from the filaments in the spinning cell. The spinning solvent is followed condensed from the cell gases, purified by distillation, and returned for next run.

A scheme of the solution dry spinning process was shown in below.

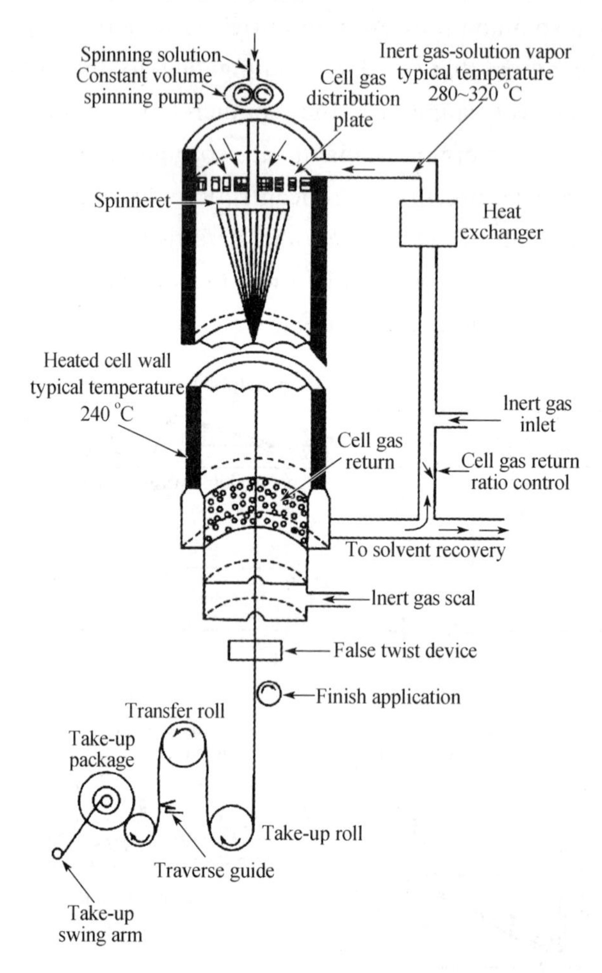

Figure 5-14 A scheme of the solution dry spinning process (F. Fourne. *Synthetic Fibers, Machines and Equipment, Manufactory, Properties.* Hanser Publisher, Munich, Germany, 1998)

5.5 Solution wet spinning

In solution wet spinning process, the spinning solution is initially pumped by a precision gear pump through spinnerettes into a solvent-water coagulation bath. Then, the same process was done as the same as in the dry spinning to form the filament. At the exit of the coagulation bath, filaments are collected in bundles of the desired tex, and a false twist was usually employed at the bath exit to give the multifilament bundles a more rounded cross section. After the coagulation bath, the multifilament bundles are counter-currently washed in successive extraction baths to remove residual solvent, then dried and heat-relaxed, generally on heated cans.

For this process, the spinning solvent is generally recovered by a two-stage process in which the excess water is initially removed by distillation followed by transfer of crude solvent to a second column where it is distilled and transferred for reuse in coming fiber manufacture.

A description of the spinning process was schemed in below (Figure 5-15).

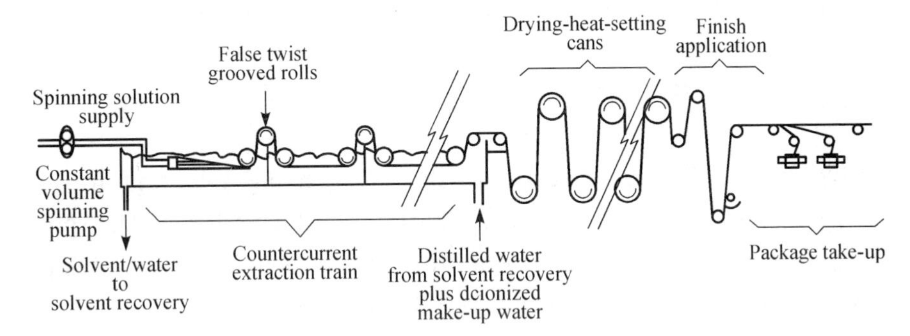

Figure 5-15 A scheme of the solution wet spinning process (F. Fourne. *Synthetic Fibers, Machines and Equipment, Manufactory, Properties.* Hanser Publisher, Munich, Germany, 1998)

5.6 Gel spinning

The main steps in the gel spinning process are described in below (Figure 5-16).

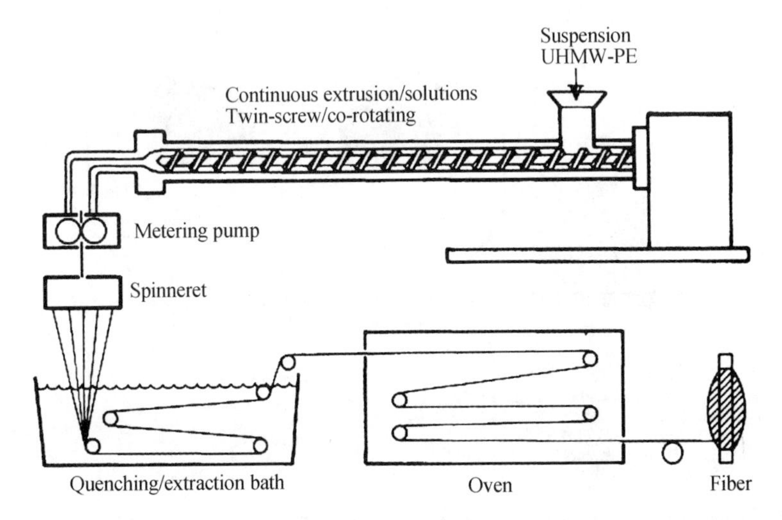

Figure 5-16 Process scheme for the production of high tenacity gel filaments (www.mdpi.com)

(1) the continuous extrusion of a solution of ultra high-molecular weight

polyethylene (UHMW-PE).

(2) Spinning of the solution, gelation and crystallization of the UHMW-PE. This can be done either by cooling and extraction or by evaporation of the solvent.

(3) Superdrawing and removal of the remaining solvent gives the fiber its final properties but the other steps are essential in the production of a fiber with good characteristics.

In the gel-spinning process, not only do all the starting parameters have an influence on the final properties of the fiber, the different process steps also influence all the following stages in the production of the fiber. So, starting from the same principles, *Dyneema* and *Spectra* may use very different equipment to produce comparable fibers.

5.7 Liquid crystal spinning

Polymer spinning solutions are extruded through spinning holes and are subjected to elongation stretch across a small air gap, as illustrated in Figure 5-17. The spinning holes fulfill an important function. Under shear, the crystal domains become elongated and orientated in the direction of the deformation. Once in the air gap, elongation stretching takes place. This is effected by making the velocity of the fiber as it leaves the coagulating bath higher than the velocity of the polymer as it emerges from the spinning holes. This ratio is often referred to as the "draw ratio" and can be fine-tuned to obtain higher tenacities and moduli with lower elongations and denier. The resulting stretch in the air gap further perfects the respective alignment of the liquid crystal domains. Overall, a higher polymer orientation in the coagulation medium corresponds to higher mechanical properties

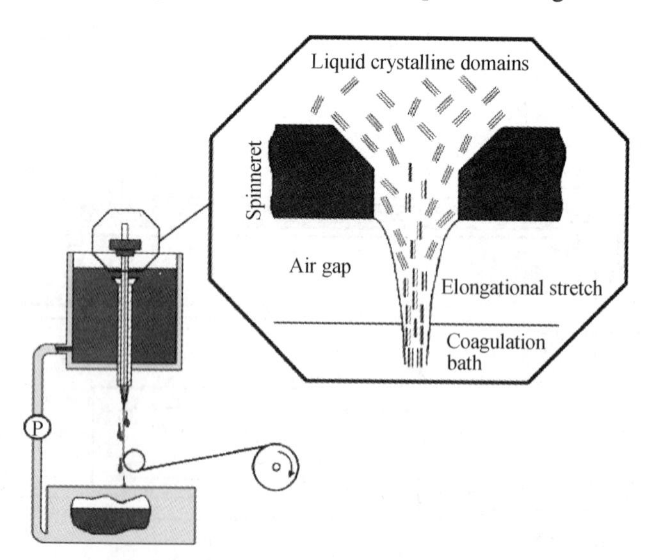

Figure 5-17 Schematically representation of the extrusion of the liquid crystalline solution in the dry-jet wet-spinning process (Franz Fourne. *Synthetic Fibers, Machines and Equipment, Manufactory, Properties.* Hanser Publisher, Munich, Germany, 1998)

of the fiber. Because of the slower relaxation time of these liquid crystal systems, the high as-spun fiber orientation can be attained and retained via coagulation with cold water. Essentially, the crystallinity and orientation of the solution are translated to the fibre. These factors allow the production of high strength, high modulus, as-spun fibers. Fibers can exhibit three possible lateral or transverse crystalline arrangements.

5.8 Electrospinning

5.8.1 General electrospinning

Electrospinning is an interesting technique for spinning polymers and its main process could be seen from Figure 5-18. Generally, this spinning method can offer fiber with excellent opportunity for designing its surface morphology and porosity to provide the most appropriate interface for biomedical applications. This process can be also regarded as a variation of the electrospraying. During the electrospinning, the solution viscosity, surface tension, conductivity, applied voltage and current are of importance. Polymer from a solution or melt can be deposited as fibrous material by charging the liquid, applying 5～30 kV. and ejecting it through a nozzle onto an oppositely charged grounded target as Figure 2-18 described. Basically, an electrospinning system consists of a high-voltage DC supply, a grounded electrode, a nozzle system with diameter controls, and a fixed or rotated target to which the spun fiber could be adhered.

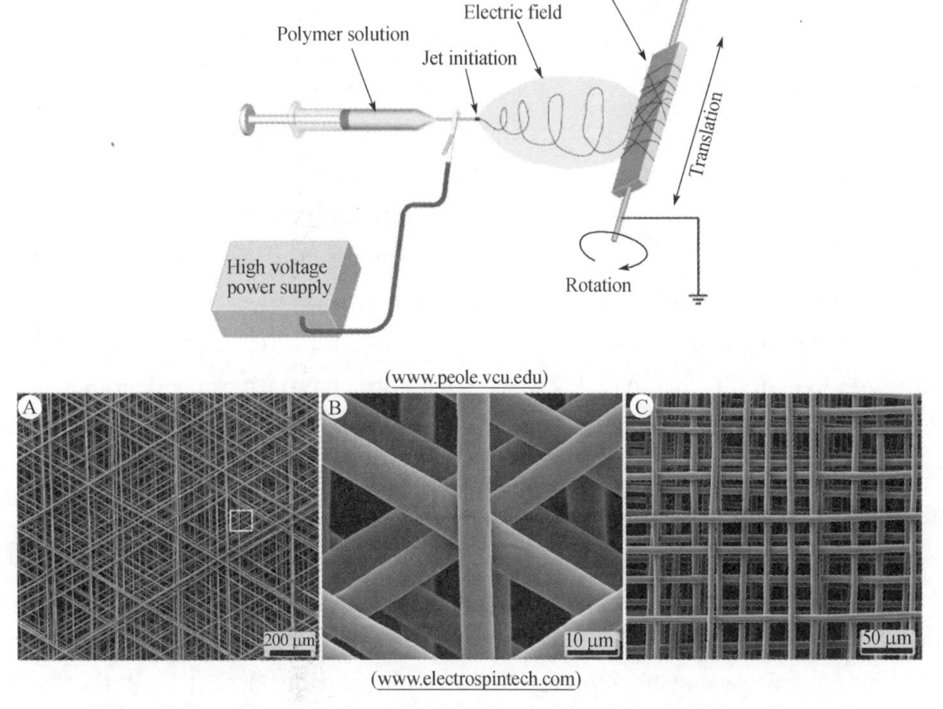

Figure 5-18 Scheme of the electrospinning (top) and formed fibers (bottom)

The electrospinning is a simple way for producing nano-size filament by properly controlling the polymer concentration and/or surface tension of the solution. The nozzle diameter can be in the range of 0. 05 μm to a few microns. In general, the fibers were spun onto nonwoven structures which are porous and have high surface area. With respect to this property, the electrospun nonwoven structure is thus availably to provide scaffolding for tissue engineering.

5. 8. 2 Coaxial electrospinning

Coaxial electrospinning is a special case of common electrospinning, and usually for producing tube-like fibers.

There have some different method for producing such products. The first is based on use of the electrospun nanofibers as templates. In this case polymeric nanofibers are produced by electrospinning and then coated with a precursor material from which the tubes are made by various deposition methods. Subsequently, the inner electrospun fiber is removed by selective dissolution or thermal degradation, and tubes with nanometric and controlled inner diameter are obtained. This method was first introduced by *Bognitzki* et al. under the name of the TUFT (tubes by fiber templates) process, to produce polymeric, polymer-metal hybrid, and metal nanotubes. *Caruso* et al. modified this process by using the sol-gel procedure to coat the template nanofiber. By using the sol-gel procedure the morphology of the template fiber, which is controlled by the electrospinning process, can be mimicked by the wall material. In this way they fabricated titanium dioxide tubes with special morphologies.

The second approach uses the co-electrospinning process, in which two different solutions are spun simultaneously using a spinneret with two coaxial capillaries to produce core/shell nanofibers (Figure 5-19). The core is then selectively removed and hollow fibers are formed. This method was used to fabricate ceramic hollow fibers by co-electrospinning viscous mineral oil as the core and a mixture of polyvinylpyrrolidone, PVP, and $Ti(OiPr)_4$ in ethanol as the shell. The mineral oil was subsequently extracted and finally, after calcination, hollow fibers made of titania were obtained. Turbostratic carbon nanotubes were also obtained by coelectrospinning of polyacrylonitrile, PAN, and poly(methyl methacrylate), PMMA, with subsequent thermal degradation of the PMMA core and finally carbonization of the PAN shell. This approach strongly depends on the stability of the co-electrospinning step, which is affected by many factors. These include the miscibility or immiscibility of the pair of solutions, viscosity ratio, viscoelastic relaxation time ratio, relative permittivity and conductivity ratios, interfacial tension, electric field strength, and degree of protrusion of the core nozzle outside of the shell nozzle. Based on experience, it seems that a well-stabilized co-electrospinning process can be achieved when both solutions are sufficiently viscous and even spinnable (at least the shell solution) and the solutions are immiscible. Although core/shell nanofibers made of miscible solutions can be achieved, this process is less controllable since mutual diffusion can take place in the Taylor cone and during jet stretching. Recently, another approach was introduced in which the co-electrospinning of a blend of two polymers through a single nozzle resulted in core/shell fibers, which were converted into carbon nanotubes.

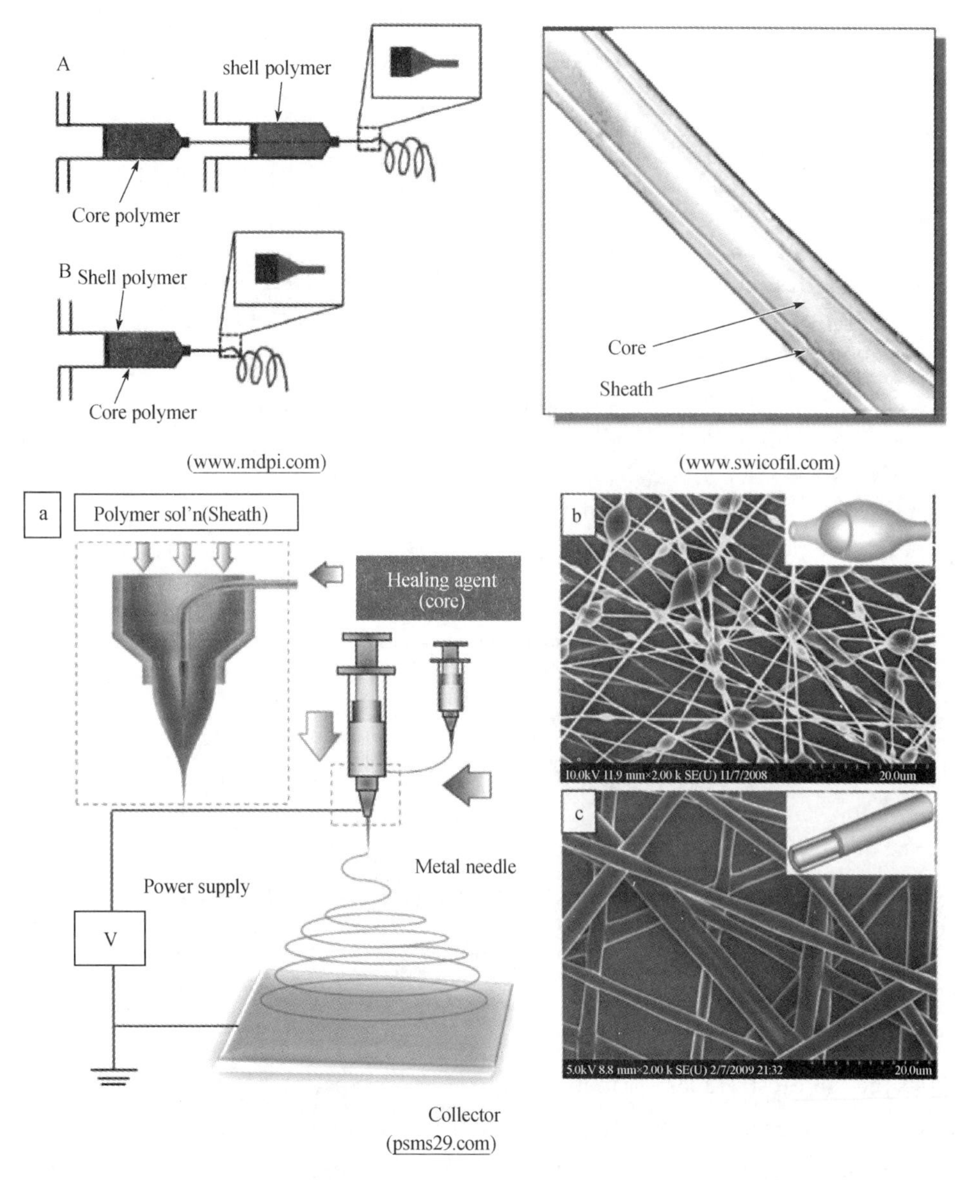

Figure 5-19 Scheme on coaxial electrospinning and related fibers

So far, all of the approaches have mainly been used to produce ceramic, carbon, or metallic tubes.

5.8.3 Gas-jet/electrospinning

In this derivate, the general electrospinning method was seen added an associated set on gas-jet. It means that the fluid flow would be enhanced by gas-jetting. The scheme of this modified electrospinning was described in below (Figure 5-20), and the special spinneret was described by Figure 5-21.

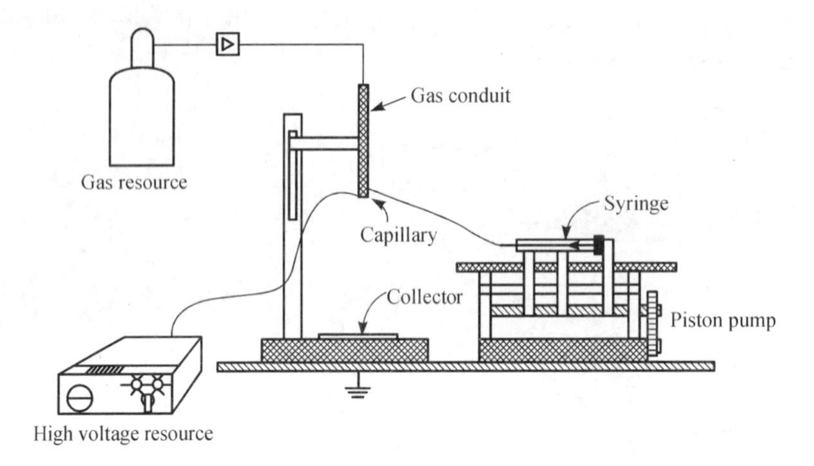

Figure 5-20 Schematic of the gas-jet/electrospinning

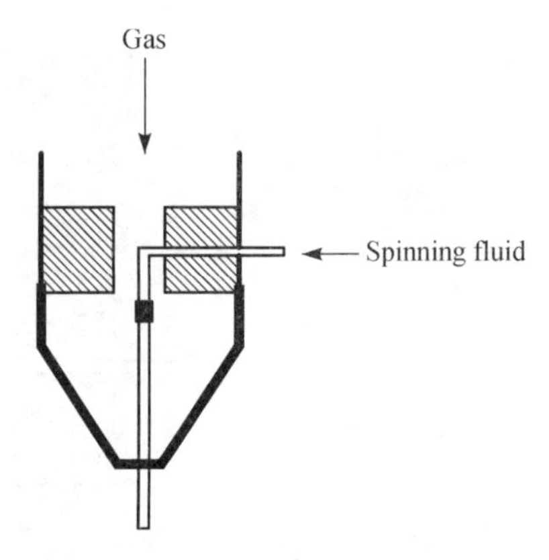

Figure 5-21 Schematic of the spinneret of the gas-jet/electrospinning

5.9 Bi-components spinning

Even the slightest traces of metals and other impurities have to be avoided in the optical core, which makes an installation according to Figure 5-22 for the production of these filaments particularly suited. The core material (2) is produced as a super pure PMMA cylinder and melted at the lower end (3) and extruded as a filament of 0.3~0.5 mm in diameter. Simultaneously it is coated by a special fluorine skin of about 12 urn thickness at (3) that comes from a second cylinder (1). After air cooling at (4) it is drawn between the godet duos (6) in a hot air oven (5) in the ratio of up to about 1:2 and wound with about 20~25 m/min. The tenacity of the filament is about 9~11 cN/tex. Optical couplers were also developed for this.

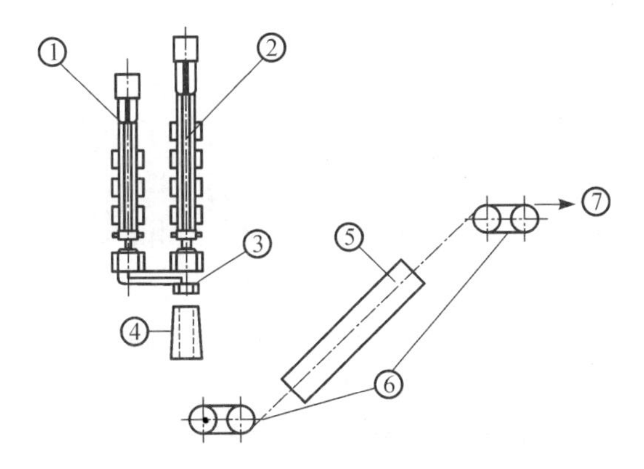

Figure 5-22 Bicomponent melt spinning installation for the production of light conductive plastic filaments. 1) Piston type extruder for the skin material; 2) Piston type extruder for the core both controlled for a synchronous quantity extrusion; 3) Bicomponent (skin-core) spinneret; 4) Air quench cooling; 5) Hot air draw duct; 6) Draw godet duo; 7) To (very large radius) winding

5.10 Reaction spinning

In the reaction spinning process the final chemical structure is not formed until after the extrusion into the coagulation bath, similar to spinning viscose. This way it is for example possible to extrude poly amide acids to filaments that are then changed by cyclohydration to polyimide filaments (Figures 5-23 and 5-24). Also some segmented polyurethane elastomer filaments can be produced by this process.

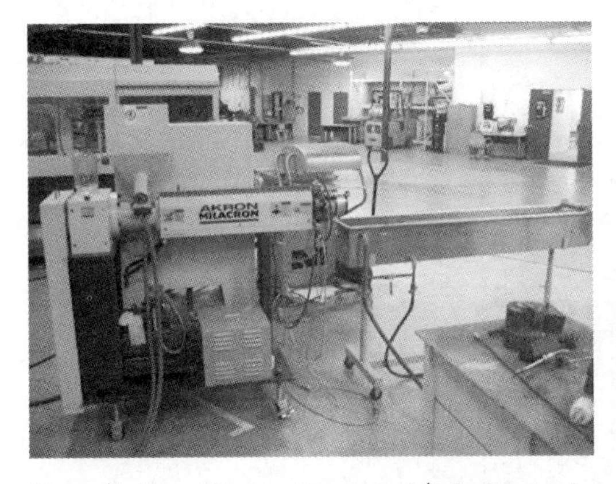

Figure 5-23 Akron milicron 1-1/2 inch extruder

Figure 5-24 Banbury external mixer

5.11 Centrifugation spinning

Poly(-3-hydroxybutyrate) fibres were produced by centrifugal spinning of the parent polymer (ex Marlborough Biopolymers Ltd, UK) from chloroform or dichloromethane. Two grades of polymer were used: a technical grade and a more rigorously purified grade. The technical grade contained *ca*. 5% of process and precursor impurities predominantly carbohydrates, which were effectively absent in the purified grade. The use of these two grades provides a useful basis for assessing the effect of particulate additives on the degradation process. Biochemicals for the determination of the monomer were in kit form from Sigma Diagnostics. Filters (0.3 mm and 3 mm) were from Millipore. Unless specified otherwise, and all chemicals from Sigma Aldrich.

Of this process, the centrifugal gel-spun PHB fibres were subjected to hydrolytic degradation in an accelerated model of pH 10.6 and 70 ℃, as described in Figure 5-25. and obtained fiber was showed in Figure 5-26.

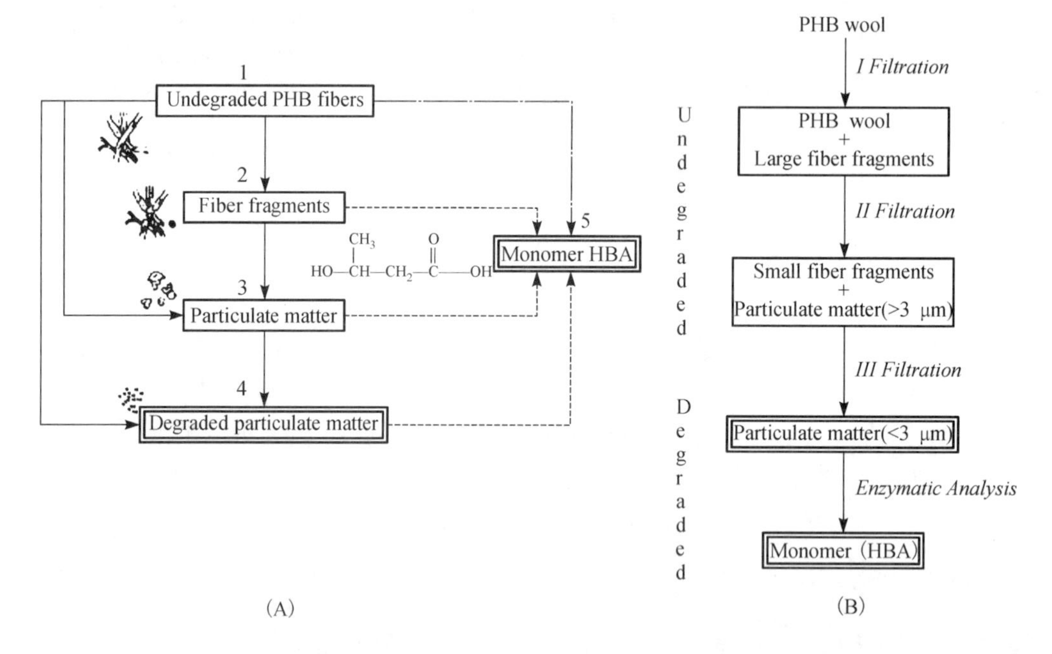

Figure 5-25 Schematic representation of the progress of degradation of PHB wool. (A) fiber fracturing and fragmentation, erosion to monomer; and (B) of the classification used to distinguish undegraded and degraded materials

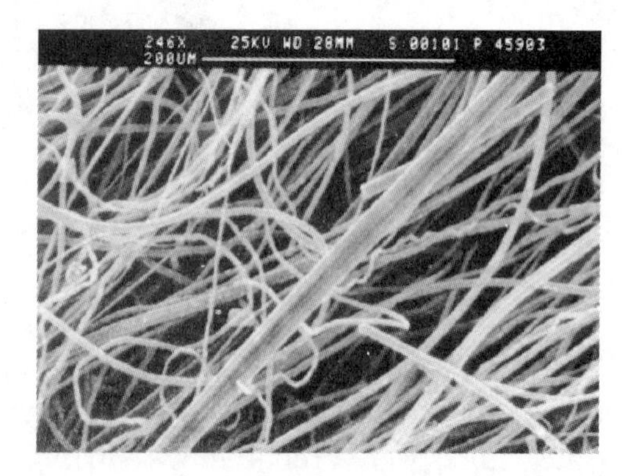

Figure 5-26 Scanning electron micrograph of centrifugally gel-spun PHB wool illustrating the dense fibrous nature of the material (1 bar=200 mm)

5.12 Interfacial polycondensation spinning

Between the solutions of dicarbon acid chloride and diamines one can produce a polycondensation in the interfacial surface of for example non-melting polyamides. Both components are dissolved in solvents that do not or only partially mix, e.g. one in water and the other in benzene. Then both are brought together in an extremely thin layer, as shown in Figure 5-27, and the formed filament or film is taken up to the top. Subsequent steps are possibly neutralization, washing, drying, preparation, and winding. This spinning process can in its most simple form be shown in a beaker. The process is a kind of reaction spinning.

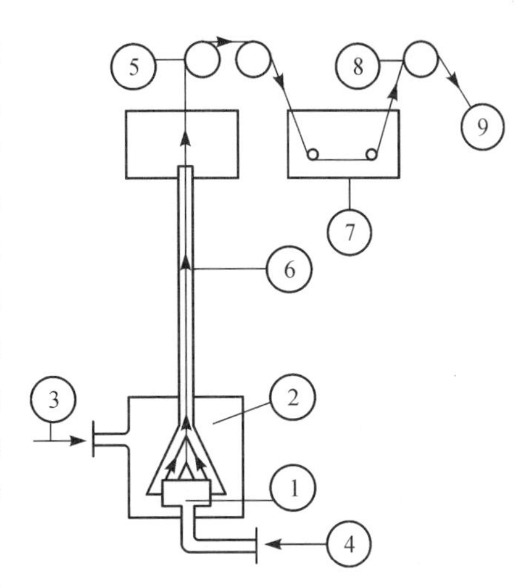

Figure 5-27 Schematic diagram of a filament spin process for interfacial polycondensation. 1, 2) Spinneret and spinning head; 3) Supply of diamine solution; 4) Supply of di-carbon acid di-chloride solution; 5) Godet take-up duo; 6) Filament formed in the boundary surface between the two solutions; 7) Washing bath; 8) Draw godet; and 9) Filament to aftertreatment

5.13 Pseudo-dry-spinning

Laure Notin et al. have reported a case on this fiber spinning method. According to these researchers, a pseudo-dry-spinning process of chitosan fiber without using any organic solvent or cross-linking agent is available. A highly deacetylated chitosan (degree of acetylation =2.7%) from squid-pens with a high weight-average molecular weight (M_w = 540 000 g/mol) was used and dissolved in acetic acid aqueous solution at a concentration of 2.4% (w/w) with a stoichiometric protonation of the —NH_2 sites. The coagulation method consisted of subjecting the extruded monofilament to gaseous ammonia. The alkaline coagulation bath classically used in a wet-spinning process was therefore not useful. A second innovation dealt with the absence of any aqueous washing bath after coagulation. The gaseous coagulation was then directly followed by a drying step under hot air. When the chitosan monofilament coagulated in the presence of ammonia gas, ammonium acetate produced with the fiber could be hydrolyzed into acetic acid and ammonia, easily eliminated in their gaseous form during drying. The pseudo-dry-spinning process did not give rise to any strong degradation of polymer chains. After 2 months at ambient atmosphere, chitosan fibers could then be stored without any significant decrease in the M_w, which remained at a rather high value of 350 000 g/mol. The obtained chitosan fibers showed a smooth, regular and uniformly striated surface.

5.14 Coaxial spinning

Coaxial spinning can be in melt or electro-methods for fabrication of nanostructured advanced materials. Usually, it is also named as the co-electrospinning for latter. In contrast to other multi-step template based procedures, this methodology is much more simple and general since, firstly, a solid template is needless and, secondly, the process is seldom affected by the chemistry of the liquids. This gentle process allows selecting the liquid precursors depending on the application sought for the nanofibers. Here, we review different products obtained by this technique:

(1) solid and hollow carbon nanofibers from different precursors (polyacrylonitrile, polyvinylpyrrolidone and lignin);

(2) nanofibers of biocompatible polymers encapsulating liquids in the form of beads;

(3) spinning nanofibers of alginate; and

(4) in-fiber encapsulation of active microgels.

5.15 Dry spinning

The dry spinning process is considerably more simples than the wet spinning process with aftertreatment (Figure 5-28). Today more than 80% of the spandex production is dry spun.

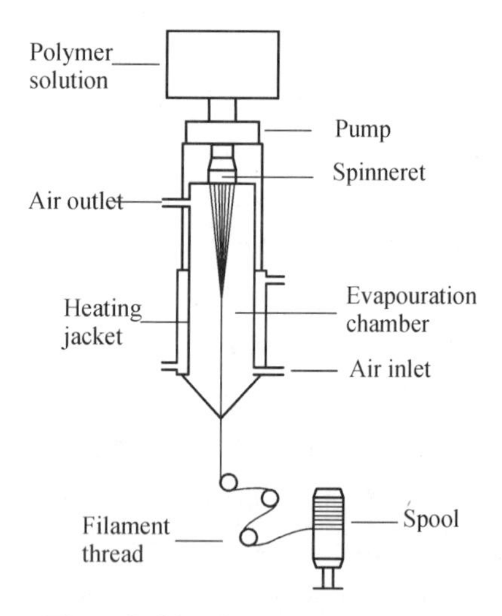

Figure 5-28 Scheme on dry spinning (www. tikp. com)

By heating solution of polymer in solvent then extruded into a hot stream of gas, which can be also continuously heated during its passage, and evaporating the solvent, the fiber therefore reacts to the loss of solvent by passing through a gel to a solid state. In such process, the fiber surface hardens initially while the solvent still diffusing out from the interior and a cylinder under external over-pressure is formed. The internal pressure decreased with the distance increased from the spinneret and the surface suddenly collapsed to rise to a round, oval or serrated, deformed cross-section, depending on the polymer. Of the dry spinning process, the exact spinning conditions, e. g. concentration, viscosity, temperature, spinning gas flow rate etc. , are polymer-and solvent specific.

On the use of the dry-spinning technique, the recovery of the solvent is absolutely essential, while the recovery of the spinning gas might be also advisably.

5.16 Dry jet-wet spinning

In wet spinning, the as-spun fiber always has voids that cause deterioration of the weak mechanical properties for fiber; probably the fibrillation and low transparency occurred easily. Since the formation of the voids can be controlled in a minimum by extruding the dope stream from a dry jet, and subsequently followed by coagulation as in conventional wet spinning, such process is regarded as the dry jet-wet spinning. The dry-jet-wet spinning can allow the stress relaxation for polymer chains in the air gap of the orientation produced in the spinneret to cause the spun fiber less oriented and more uniform than that from the immersed jet. This permits orientation by subsequent drawing and gives fibers with higher tenacity. This novel fiber spinning technique was schemed in Figure 5-29.

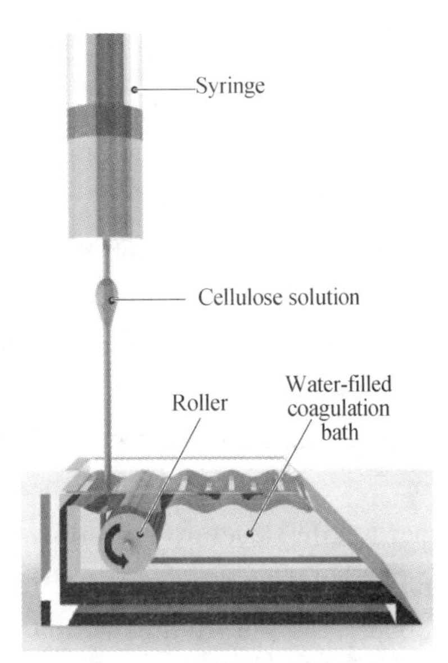

Figure 5-29 Scheme on dry-jet-wet spinning (www.mit.edu)

5.17 Emulsion and suspension spinning

According to this process filaments can be produced from non-soluble and non-melting polymers. First they are dispersed in a commonly spinnable mass, possibly with the aid of a spinning agent. This dispersion is spun and treated following the process for the carrier substance, and the latter is then dissolved, melted, evaporated or pyrolized. The difficult part is to keep the dispersed mass together during the removal of the carrier. Size and form of the suspended parts are decisive for the coherence, the tenacity of the filament, etc. Parts of the size 5~1 000 nm at a ratio of diameter to thickness of about 400 : 1 are advantageous. Thus PTFE and others can be transformed into filaments, also mixtures of PVC with PVA or matrix fibril filaments from PA 6 and PET or ceramic that was spun and aftertreated in viscose solutions.

5.18 Inviscid melt spinning

A eutectic composition (46.5% CaO, 53.5% Al_2O_3) calcium aluminate fiber formed by Inviscid Melt Spinning (IMS) was recently reported by *Brian S. Mitchell* who analyzed obtained fiber using micro-Raman spectroscopy. The fiber surface was shown to consist of pyrolytic graphite. Analysis along the fiber cross-section showed that the pyrolytic graphite layer had a total thickness of approximately 15 μm. Micro-Raman analysis of the fiber bulk was consistent with amorphous calcium aluminate ($CaAl_2O_4$).

5.19 Aqueous sol-gel blow spinning

Rajendran and Bhattacharya recently reported a case on this method. Rare-earth orthoferrite, $LnFeO_3$ (Ln = La, Sm, Gd, Dy, Er and Yb) ceramic fibers were produced by aqueous sol-gel blow spinning process at low-temperatures. Stable, charge stabilized, colloidal precursor sols of orthoferrites were prepared by room temperature processing of inexpensive and commercially available metal salts. The average diameter (Z_{av}) of the colloidal sol particles was in the range of 4～7 nm and had a narrow size distribution. The sols were concentrated, combined with spinning aids, and processed further to a viscous "spinning solution". The gel fibers of about 6 μm diameter were blow spun, collected as random fibers, dried and heated to increasingly higher temperatures at a rate of 50 ℃/h. The gel fibers converted to flexible ceramic fibers, and single-phase perovskite structure crystallized directly for all the $LnFeO_3$ (Ln = La, Sm, Gd, Dy, Er and Yb) fibres on heating them to 700 ℃. The ceramic fibers had mean diameter of about 3～4 μm, and consisted of randomly oriented submicron sized grains.

5.20 Planar-flow melt spinning

Byrne et al. reported a novel spinning process which is named as the planar-flow melt spinning (PFMS) and a single stage rapid manufacturing/solidification technique for producing thin metallic sheet or foil. A new technology, envisioned to allow real-time manipulation of the local cooling rates and properties in melt-spun ribbon, has been tested successfully when casting Al-7% Si. Pulsed laser heating, directed low on the upstream meniscus, or on the substrate, leaves patterns of "dimples" in the ribbon. Typical cooling rates of 10^4 K/s have been measured using a control-volume approach. Secondary dendrite arm spacing (SDAS) has been measured through the thickness of ribbons showing areas both affected by the laser heating and unaffected by the laser. Through a correlation of cooling rates and SDAS, it is shown that the unmodified ribbon has an average cooling rate similar to that measured macroscopically. The cooling rate underneath a laser dimple is estimated to be six times slower near the contact surface. It is envisioned that the technology described may bring the concept of "casting-by-design" one step closer to realization.

5.21 Self-assembly spinning

Most natural fibers (animal and plant) are formed using the self-assembly spinning method. In addition, a lot of micro-/nano-structure fibers directly formed in polymerization process which can be also regarded as the self-assembly spun.

5.22 Electro-assembly spinning

Unlike the self-assembly spinning, the electro-assembly spinning is an electricity-forced assembly process which could be occurred in polymerization process to direct form nanofiber. Of such process, the fiber formation was strongly controlled by applied electricity. Since such case usually applying an electrostatic generator to assist the assembly and this case could provide two electric cycles in relation to the electrostatic force enhance or reduce due to the positive and negative electrodes alternatively linked to reaction media (also in charge) and reactor (in metal).

For ample, in the electro-solution polymerization of PANI nanofibers using a device as Figure 5-30 described, we obtained longer and shorter PANI nanofibers as presented in Figure 5-31.

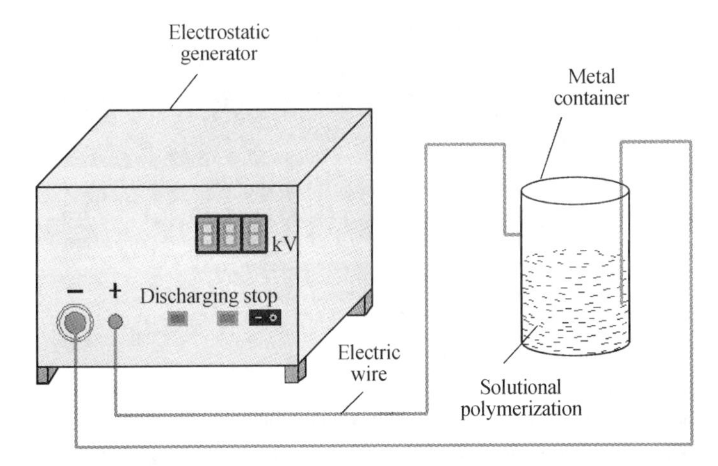

Figure 5 - 30 **Scheme on the electro-synthesis of PANI nanofibers. In experiment, the positive and negative electrodes were alternatively linked to either the ANI solution or the metal container to form two electric cycles corresponding to electrostatic interactions enhance or reduce, respectively**

Synthesis condition	SEM images of PANI	Comment
Normal solution synthesis		No fiber shape

Synthesis condition	SEM images of PANI	Comment
An EI reduce-based electro-synthesis by taking the positive and negative electrodes linking to the ANI solution and metal container, respectively	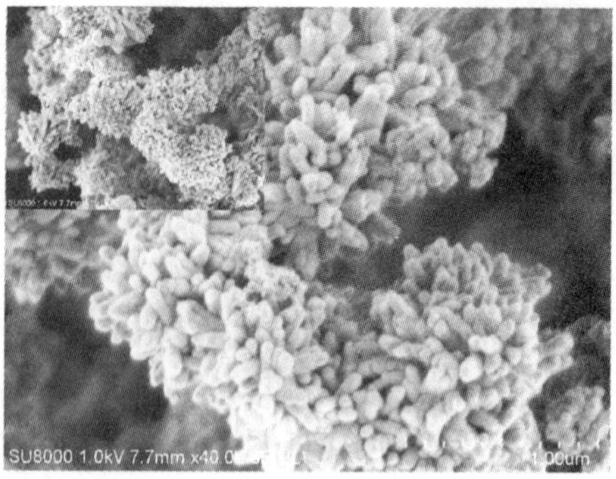	$d_a = 100$ nm $L = 200 \sim 300$ nm
An EI enhance-based electro-synthesis by taking the positive and negative electrodes linking to the metal container and ANI solution, respectively	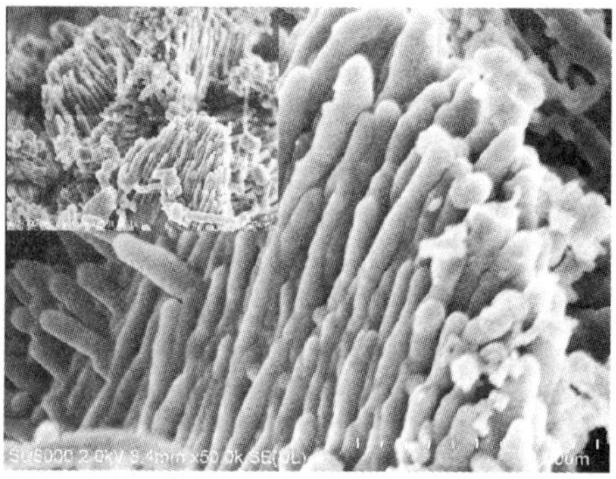	$d_a = 100$ nm $L \geqslant 1$ μm

Figure 5-31 SEM images of normal PANI sample (top) and electro-synthesized PANI nanofibers in relation to the electrostatic interactions, EI, reduce (middle) or enhance (bottom), respectively

5.23 Synthesis directly spinning

This method represents the polymerization process directly formed fiber. This is because in solvent or solution environment, some fiber structures, especially the nanofiber, can be directly formed in polymerization processes via self-assembly or template guided spun.

5.23.1 Emulsion polymerization direct spinning

During the emulsion polymerization of PANI, some cases proven that the nanofibers are availably self-assembled corresponding to the direct spun.

5.23.2 Solution polymerization direct spinning

Applying the α-, β- and γ-cyclodextrin, CD, as template or the lignosulfonate, LGS, via the solution polymerization, several polyaniline nanofibers were directly prepared (Figure 5-32) and this process can be defined as the solution polymerization-based direct spinning. This is also a self-assembly spinning process.

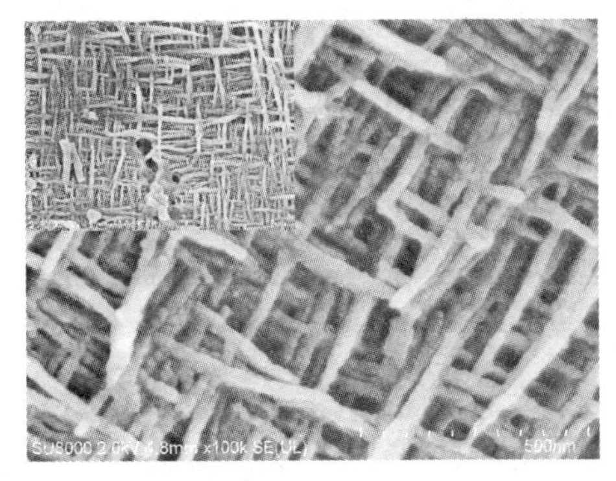

PANI nanofiber guided by β-CD

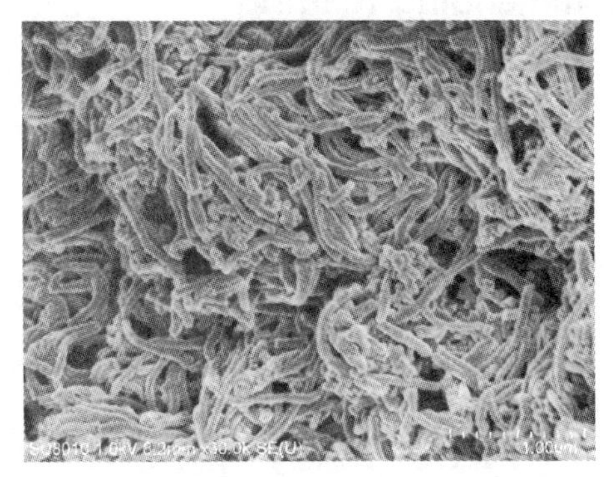

PANI nanofiber guided by LGS

Figure 5-32 SEM images of normal PANI nanofiber guided by β-CD or LGS, respectively

5.23.3 Interfacial polymerization-based direct spinning

In the case of using PLLA and PDLA as two guiders, the interfacial polymerization of PANI was performed due to the PLA dissolved in solvent formed an oil phase which different than that of the solution phase, we directly obtained two PANI fibers with different morphologies as showed in Figure 5-33.

PLLA-guided polyaniline fiber
ANI/PLLA = 344/1

PDLA-guided polyaniline fiber
ANI/PDLA = 344/1

Figure 5-33 SEM images of PANI fiber guided by PLLA and PDLA via interfacial polymerization, respectively

5.24 Indirect spinning

In general, most common spinning methods can be defined as the indirect spinning because the fiber spun from these processes from the original polymeric or related raw materials. In other words, except the fiber can be formed during the polymerization, all others are the indirect spinning process.

Recommending reading

[1] F. Fourne. *Synthetic Fibers, Machines and Equipment, Manufactory, Properties*. Hanser Publisher, Munich, Germany (translated by Dr. Helmut H.A. Hergeth, Raleigh, NC, USA, and Ron Mears, Obernburg, Germany), 1998.

[2] J. W. S. Hearle. *High-Performance Fibers*. CRC Press, Woodhead Publishing Ltd, Cambridge, England, 2001.

[3] T. Hongu, O. G. Phillips, M. Takigami. *New Millennium Fibers*. CRC press, Woodhead Publishing Limited, 2005.

[4] I. G. Loscertales, Juan E. Díaz Gómez, M. Lallave, J. M. Rosas, Jorge Bedia, J. Rodríguez-Mirasol, T. Cordero, M. Marquez, S. Shenoy, G. E. Wnek, T. Thorsen, A. Fernúndez-Nieves, A. Barrero,

[5] B. S. Mitchell. *Materials Letters*. **2000**, 45, 138-142.

[6] M. Rajendran, A. K. Bhattacharya. *J Euro Ceramic Soc*. **2004**, 24, 111-117.

[7] C. J. Byrne, A. M. Kueck, S. P. Baker, P. H. Steen. *Materials Sci Eng A*. **2007**, 459, 172-181.

[8] Z. X. Wang, Q. Shen, Q. F. Gu. *Carbohydrate Polym*, **2004**, 57, 415-418.

[9] D. L. Li, Q. Shen, H. G. Ding, Q. F. Gu, Z. X. Wang. *J. Appl Polym. Sci*. **2006**, 101, 2810-2813.

[10] Q. Yang, F. Dou, B. Liang, Q. Shen. *Carbohydrate Polym*. **2005**, 59, 205-210.

[11] Q. Yang, F. Dou, B. Liang, Q. Shen. *Carbohydrate Polym*. **2005**, 61, 393-398.

[12] Q. Shen, T. Zhang, W. X. Zhang, S. Chen, M. Mezgebe. *J Appl Polym Sci*. **2011**, 121, 989-994.

[13] J. T. Wang, L. L. Li, M. Y. Zhang, S. L. Liu, Q. Shen. *Mater Sci Eng C*. **2014**, 34, 417-421.

[14] J. T. Wang, L. L. Li, L. Feng, J. F. Li, L. H. Jiang, Q. Shen. *Int'l J Biological Macromolecules*. **2014**, 63, 205-209.

[15] J. R. Ye, S. Zhai, Z. J. Gu, N. Wang, H. Wang, Q. Shen. *Mater Lett*. **2014**, 132, 377-379.

[16] L. Notin, et al. *Acta Biomaterialia*. **2006**, 2, 297-311.

Problems

1. Please add a new fiber formation method by searching published literature elsewhere.
2. Please describe a method for directly formation of fiber by synthesis.

Chapter 6

Methods for Fiber Measurement and Characterization

Systematic investigation of the structure-property relationships for fiber materials is required and necessary because it allows the prediction of the physical properties from a knowledge of the chemical structure of the repeat units and molar mass of the fiber used polymer materials. All polymeric materials can be divided into a series of subclasses, reflecting either their method of synthesis or some particular characteristic of the material. Using this classification, it is possible to quickly identify how the material will respond to external factors such as change of temperature, pressure, stress, impact, etc. Therefore, different methods and techniques were applied to measure and characterize the formed fiber or it related materials.

6.1 Morphology characterization

6.1.1 Scanning electron microscope

The scanning electron microscope, SEM, is undoubtedly the most widely used of all electron beam instruments. The popularity of the SEM can be attributed to many factors: the versatility of its various modes of imaging, the excellent spatial resolution now achievable, the very modest requirement on sample preparation and condition, the relatively straightforward interpretation of the acquired images, the accessibility of associated spectroscopy and diffraction techniques. And most importantly its user-friendliness, high levels of automation and high throughput make it accessible to most research scientists. With the recent generation of SEM instruments, high-quality images can be obtained with an image magnification as low as about 5 × and as high as > 1 000 000 ×; this wide range of image magnifications bridges our visualization ability from naked eyes to nanometer dimensions. Image resolution of about 0.5 nm can now be achieved in the most recent generation field-emission-gun SEM (FEGSEM), clearly rivaling that of a transmission electron microscope (TEM); the sample size, however, can be as large as production-scale silicon wafers.

Some SEM images of different fibers were presented in below (Figure 6-1).

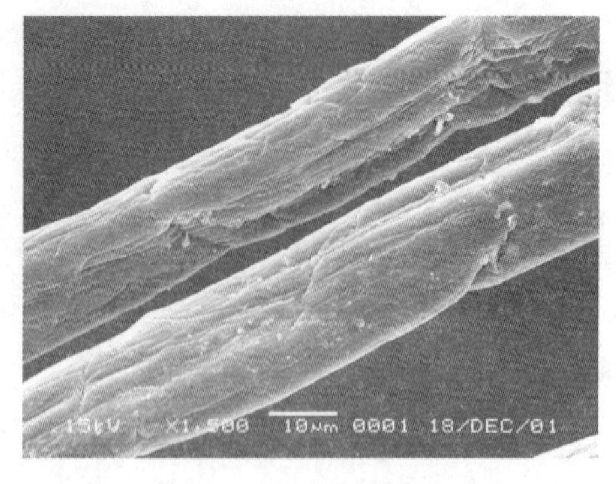

Normal chitosan fiber

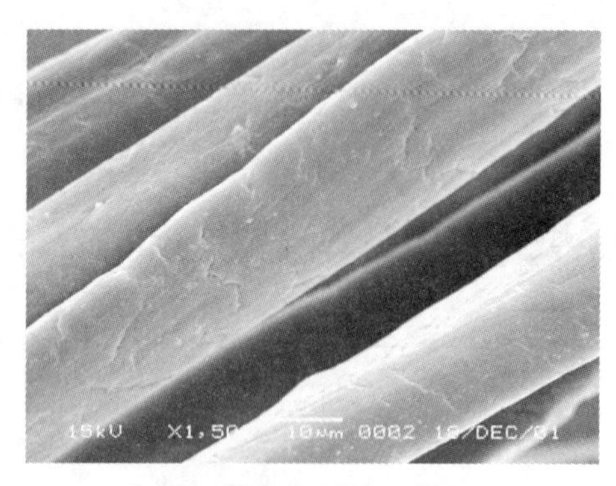

Cross-linked chitosan fiber

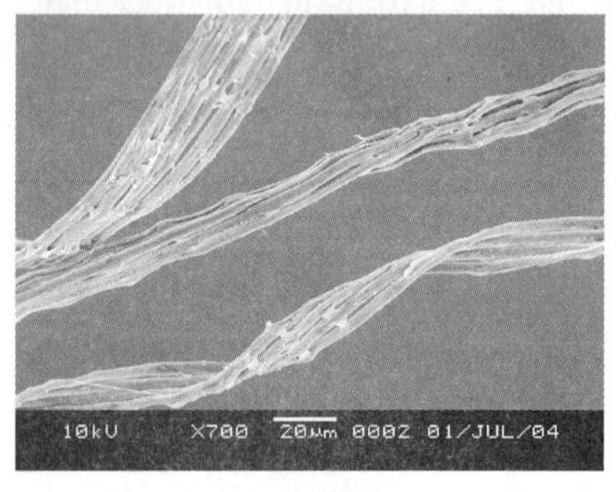

Lotus fiber

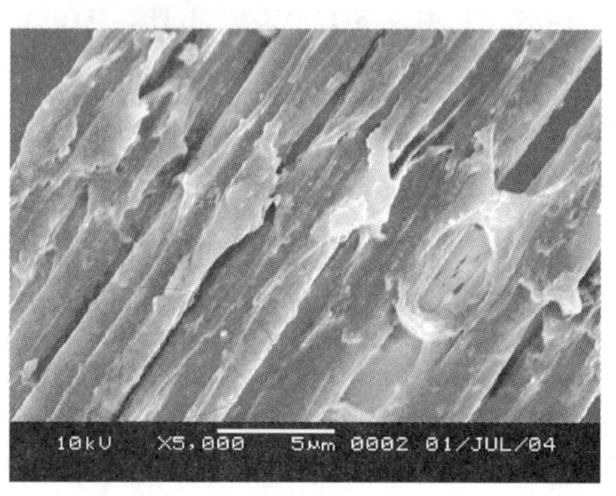

Lotus fiber

PLLA-guided polyaniline fiber ANI/PLLA = 344/1

PDLA-guided polyaniline fiber ANI/PDLA = 344/1

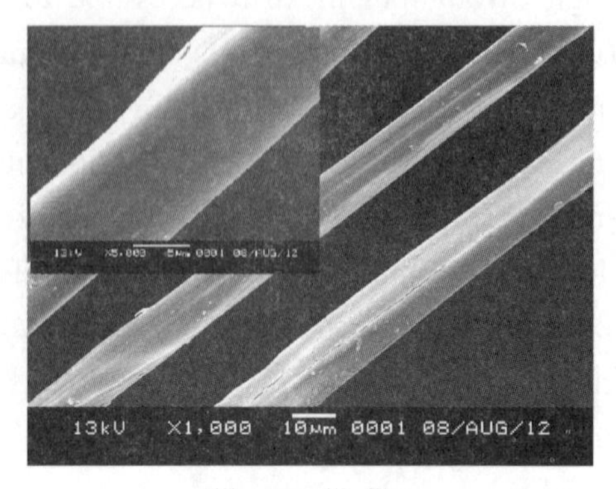

Silkworm silk fiber

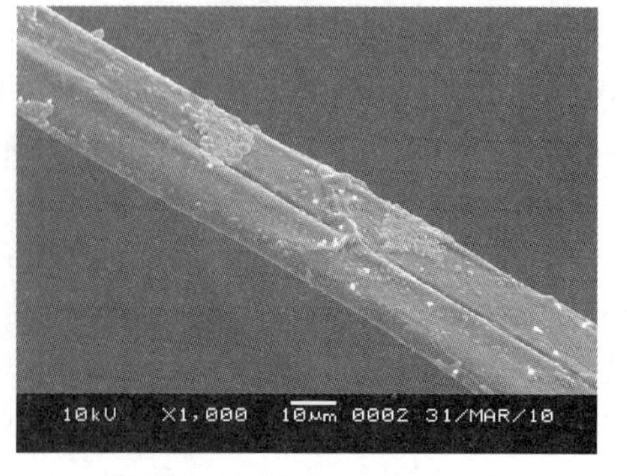

Silk fiber with fed carbon nanotube

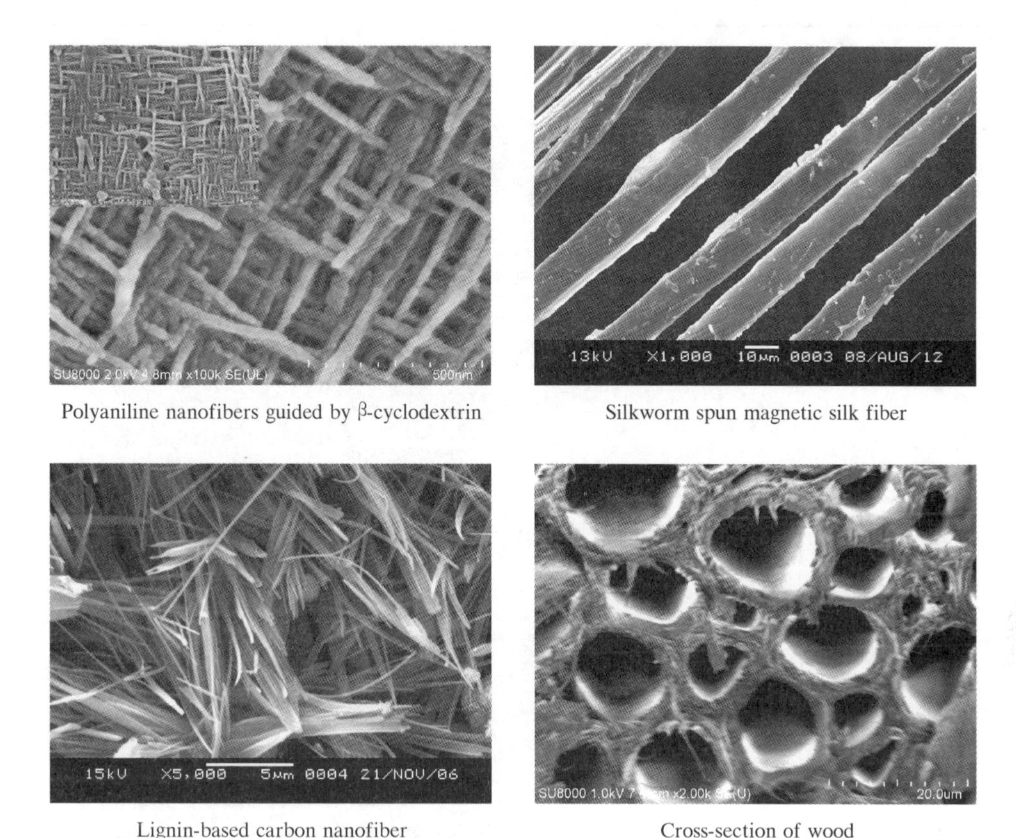

Polyaniline nanofibers guided by β-cyclodextrin

Silkworm spun magnetic silk fiber

Lignin-based carbon nanofiber

Cross-section of wood

Figure 6-1 SEM images of various fibers

6.1.2 Transmission electron microscope

The transmission electron microscope, TEM, has evolved over many years to such an extent that resolving powers at or close to the 0.1 nm level can nowadays be attained almost on a routine basis. Using correct operating conditions and well-prepared samples, TEM image characteristics are interpretable directly in terms of projections of individual atomic-column positions. With quantitative recording and suitable image processing, atomic arrangements at defects and other inhomogeneities can be reliably and accurately determined. Since nanoscale irregularities have a marked influence on bulk behavior, the TEM has become a powerful and indispensable tool for characterizing nanomaterials.

The process of image formation in the TEM can be considered as occurring in two stages. Incoming or incident electrons undergo interactions with atoms of the specimen, involving both elastic and inelastic scattering processes. The electron wave function which leaves the exit surface of the specimen is then transmitted through the objective lens and subsequent magnifying lenses of the electron microscope to form the final enlarged image.

TEM images of some PANI nanofibers (nanotubes) were showed in below (Figure 6-2).

Polyaniline nanofibers guided by β-cyclodextrin with various ratios	
Total added mass 100%	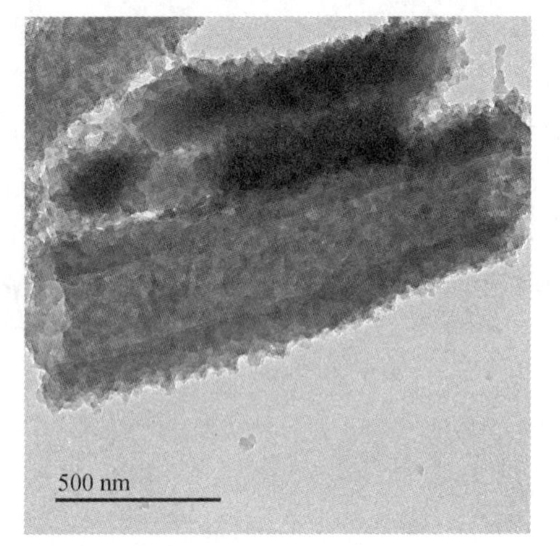
Total added mass 75%	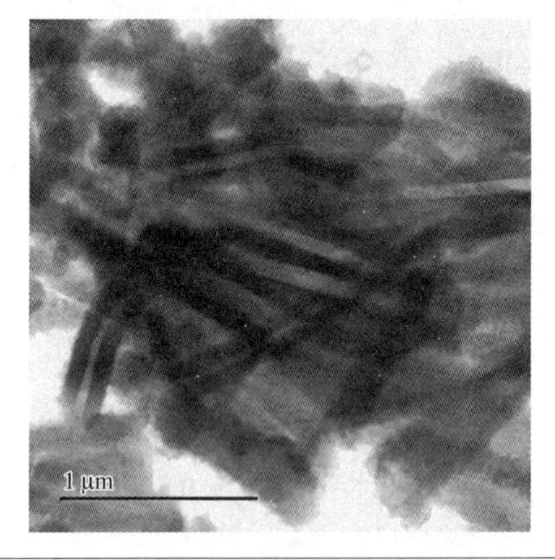
Total added mass 50%	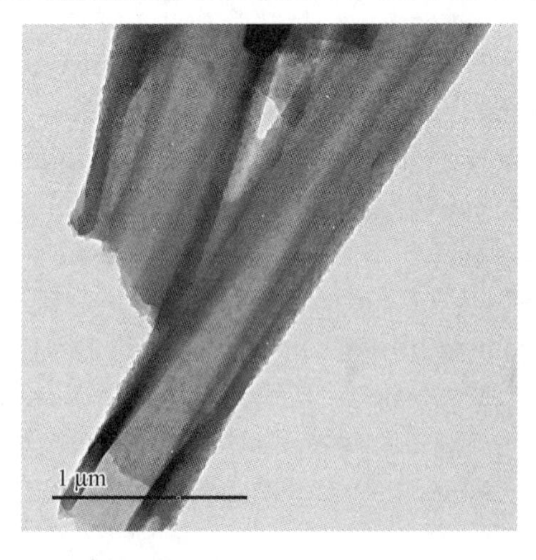

Figure 6-2　TEM images of some PANI nanotubes

6.1.3 Scanning tunneling microscope

In March 1981, Binnig, Rohrer, Gerber and Weibel at the IBM Zurich Research Laboratory successfully developed the first STM in the world by combining vacuum tunneling with scanning capability. In this first STM, a sharp metal tip fixed on the top of a pizeodrive, PZ, to control the height of the tip above a surface. When the tip is brought close enough to the sample surface, electrons can tunnel through the vacuum barrier between tip and sample. Applying a bias voltage on the sample, a tunneling current can be measured through the tip, which is extremely sensitive to the distance between the tip and the surface as discussed above. Another two pizeodrives defined as the PX and PY are used to scan the tip in two lateral dimensions. A feedback controller is employed to adjust the height of the tip to keep the tunneling current constant. During the tip scanning on the surface, the height of the tip (the voltage supplied to PZ pizeodrive) is recorded as the STM image, which represents the topograph of the surface. This operation mode of STM is called "constant current" mode.

Constant current mode is mostly used in STM topograph imaging. It is safe to use the mode on rough surfaces since the distance between the tip and sample is adjusted by the feedback circuit. On a smooth surface, it is also possible to keep the tip height constant above the surface, then, the variation of the tunneling current reflects the small atomic corrugation of the surface. The constant height mode has no fundamental difference to the constant current mode. However, the tip could be crashed if the surface corrugation is big. On the other hand, the STM can scan very fast in this mode for research of surface dynamic processes.

STM has a great problem in its measurement because it needs the sample in electric.

6.1.4 Atomic force microscope

Since the STM presented problem in request of sample in conductivity, the atomic force microscope, AFM, was furthermore developed. The scope of AFM applications includes high-resolution examination of surface topography, compositional mapping of heterogeneous samples and studies of local mechanical, electric, magnetic and thermal properties. These measurements can be performed on scales from hundreds of microns down to nanometers, and the importance of AFM, as characterization technique, is further increasing with recent developments in nanoscience and nanotechnology. In studies of surface roughness, AFM complements optical and stylus profilometers by extending a measurement range towards the sub-100 nm scale and to forces below nano-Newton. These measurements are valuable in several industries such as semiconductors, data storage, coatings, etc. AFM together with scanning electron microscopy of critical dimensions is applied for examination of deep trenches and under-cut profiles with tens and hundreds of nanometers dimensions, which are important technological profiles of semiconductor manufacturing.

AFM is capable for applying to study heterogeneous materials in industries, e.g. the synthesis, design, and applications.

The scheme of an AFM process and related force were presented in Figure 6-3 as below.

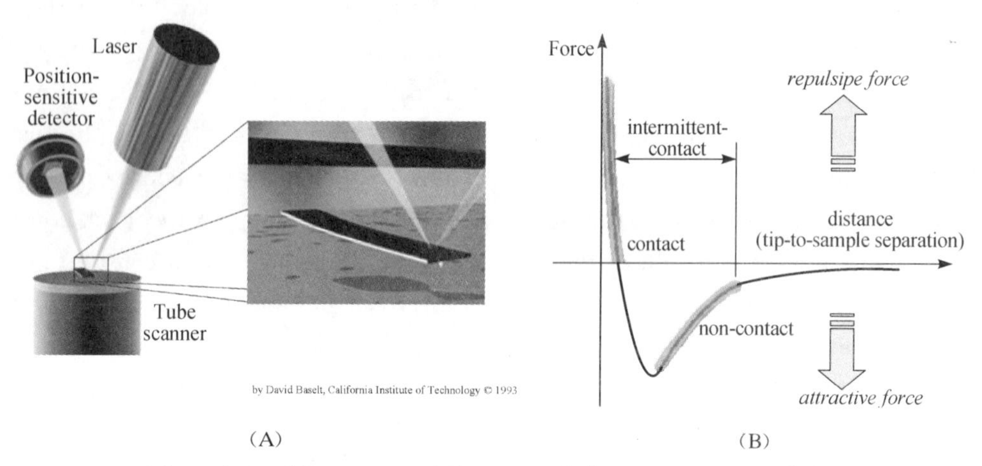

Figure 6-3　Scheme on the AFM process (A) and related forces (B)

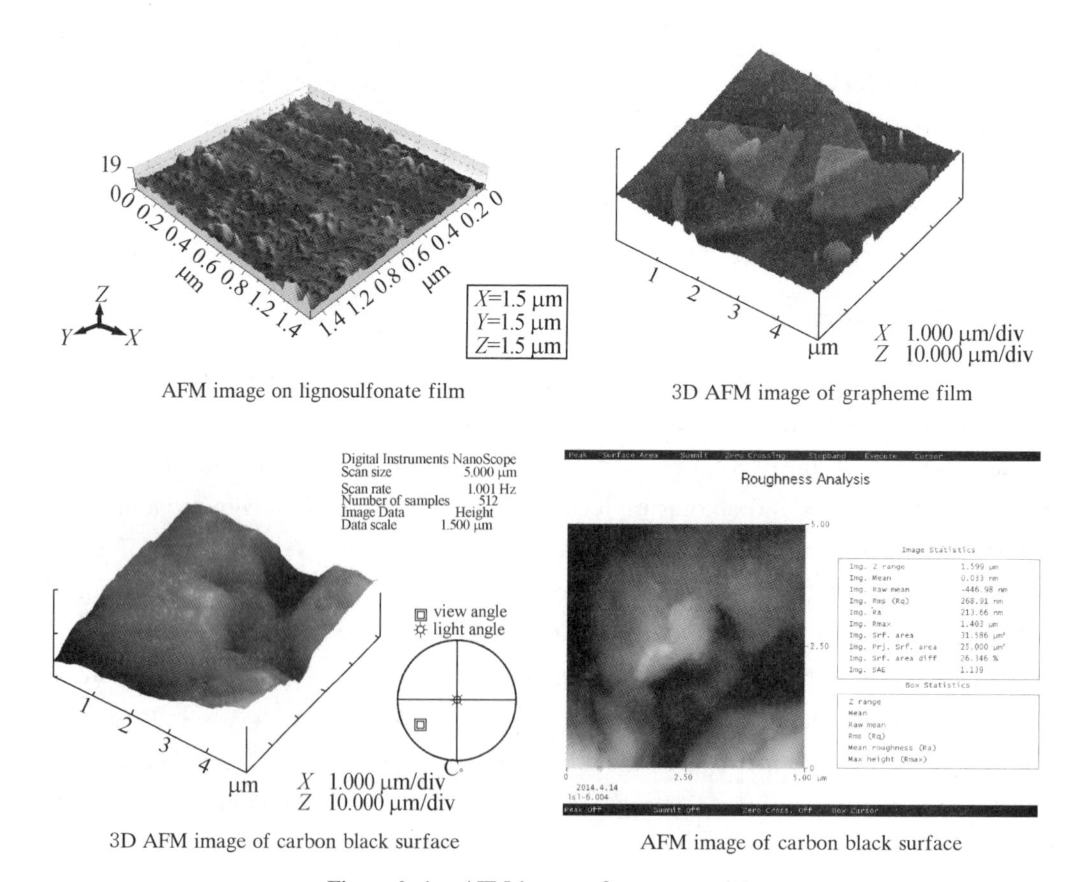

Figure 6-4　AFM images of some materials

6. 1. 5 Optical microscope

In general, optical microscopy has played an important role in the visualization of ordered polymer fibers, and a range of different approaches can be adopted to enhance the contrast, phase contrast, polarized light microscopy, orientation birefringence, strain birefringence, modulated contrast, interference microscopy, etc. Optical microscopy rarely provides resolution better than several micrometers but can give a quick and easy assessment of the extent of order in the polymer fiber.

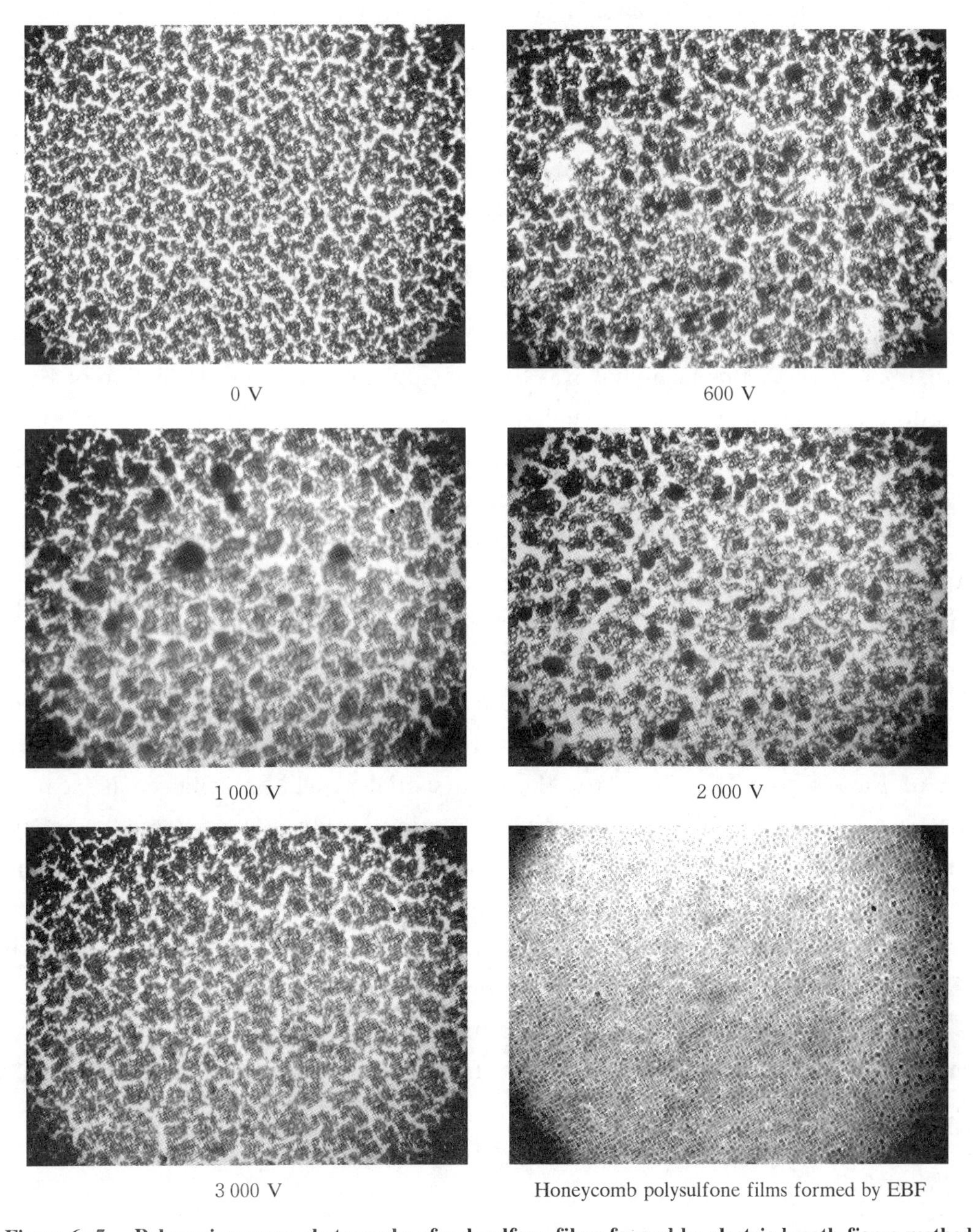

Figure 6-5 Polar microscopy photographs of polysulfone films formed by electric breath figure method

Of those various microscopes, the polarized light microscopy is broadly applied in fiber area due to its low cost and wide suitability in fiber aspects study, especially the orientation. Figure 6-6 showed some photographs of polymer films obtained from polarized light microscopy.

6.2 Structure characterization

Identification of the type of functional groups present in a fiber is effectively achieved by infrared and Raman analysis. The spectroscopic selection rules for infrared and Raman activity are respectively a dipole moment or polarizability change during interaction of electromagnetic radiation with the atomic grouping. A vibration that gives a strong infrared signature may be weak in the Raman spectrum and vice versa. It is relatively easy to identify the occurrence of carbonyl, ethers, aromatic functions, hydroxyl groups, epoxy rings, carbon-halogen, and carbon-hydrogen bonds.

6.2.1 Fourier transform Raman spectroscopy

The discovery of Raman scattering is by Krishna and Raman in 1928. Until approximately 1986, this technique was broadly applied with the introduction of Fourier transform (FT)-Raman, charge-coupled devices, small computers, and near-infrared lasers. These developments overcame the major impediments and resulted in a Raman renaissance in the context of detailed applications.

Raman spectroscopy is a promising tool combined with the near-field techniques and available for identifying and analyzing the molecular composition of complex materials. Vibrational spectra directly reflect the chemical composition and molecular structure of a sample. By raster scanning the sample and point wise detection of the Raman spectra, chemical maps with nanoscale resolution can be obtained. A main drawback of Raman methods is the low scattering cross-section, typically 14 orders of magnitude smaller than those of fluorescence. Furthermore, the high spatial resolution achieved in near-field optics is linked to tiny detection volumes containing only a very limited number of Raman scatterers. The weakness of Raman signals in combination with the limited transmission of typical aperture probes requires extended integration times even intermediate aperture sizes of about 150 nm. An essential improvement of the spatial resolution below 50 nm with smaller apertures appears to be unfeasible.

The signal enhancement achieved in tip-enhanced Raman spectroscopy can be demonstrated by comparing the Raman spectra detected in presence and absence of the enhancing metal tip.

Figure 6-6 showed FT-Raman spectra of some materials.

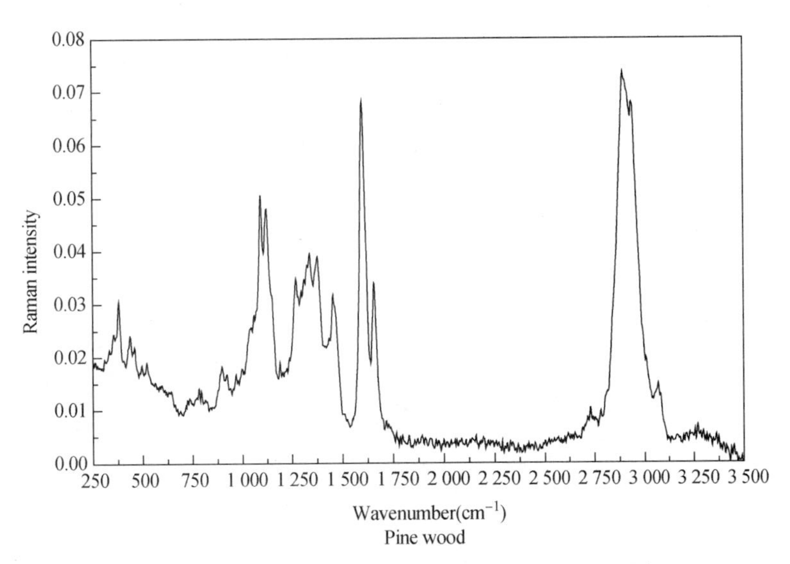

Pine wood

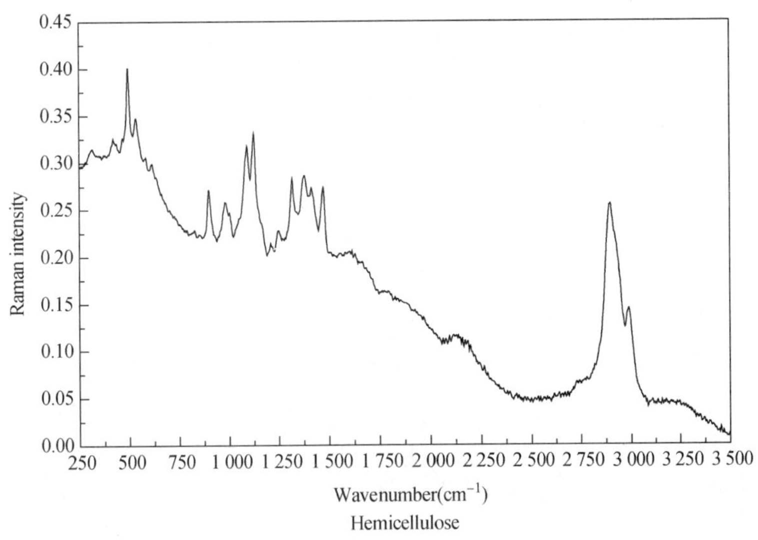

Hemicellulose

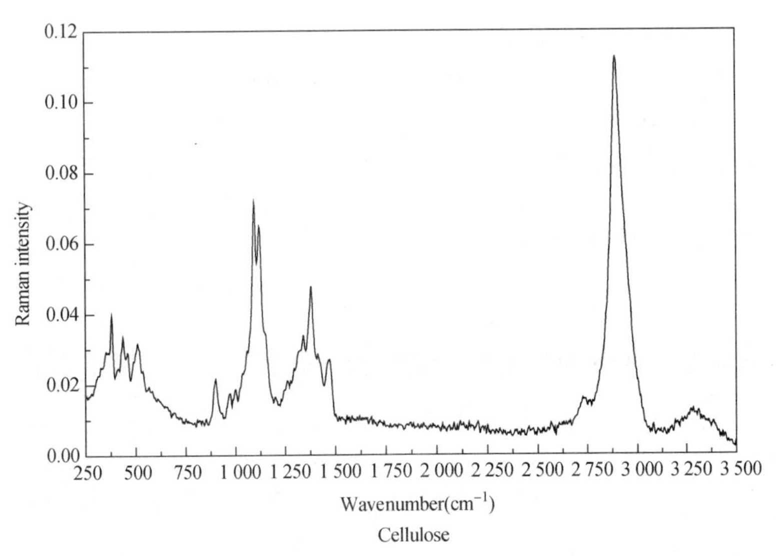

Cellulose

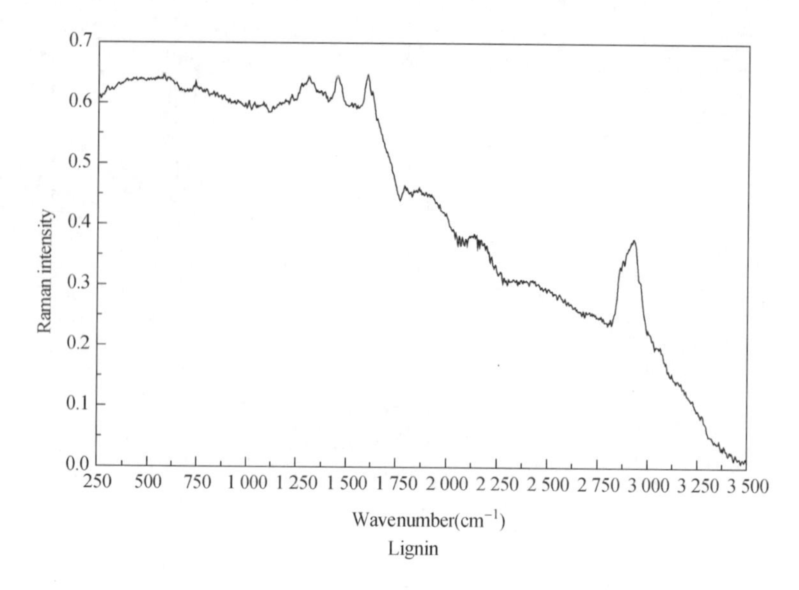

Figure 6-6 FT-Raman spectra of some materials

6.2.2 Fourier transform infrared spectroscopy

Fourier transform infrared spectrometer, FTIR, can provide absorption spectra and is among the oldest and most widely used characterization techniques available to a lot of materials. Sample preparation does not require significant effort. Polymers mixed with potassium bromide and then pressed into pellets or films prepared from melt or cast from solution can be easily studied. For bulk samples or powders, or if a concentration profile is needed for a film, the reflectance technique is perhaps more suitable.

Normal vibrations related to a change in dipole moment are infrared active. Groups with large dipole moments, such as C=O and N—H, typically have strong infrared absorptions. The majority of reported spectroscopic studies by infrared spectroscopy focus on determination of polymer molecular composition by analysis of characteristic vibrations of functional groups. The power of vibrational spectroscopy, i. e. its selectivity and sensitivity, cannot be overestimated. With accurately defined band assignment, particularly if the transition dipoles are well established, quantitative analysis of sample anisotropy in terms of segmental orientation can be accurately established.

Below showed some FTIR spectra of materials (Figure 6-7).

6.2.3 Nuclear magnetic resonance spectroscopy

Nuclear magnetic resonance, NMR, spectroscopy is a well-known and popular technique for material characterization. Subatomic particles (electrons, protons and neutrons) can be imagined as spinning on their axes. In many atoms (such as ^{12}C) these spins are paired against each other, such that the nucleus of the atom has no overall spin.

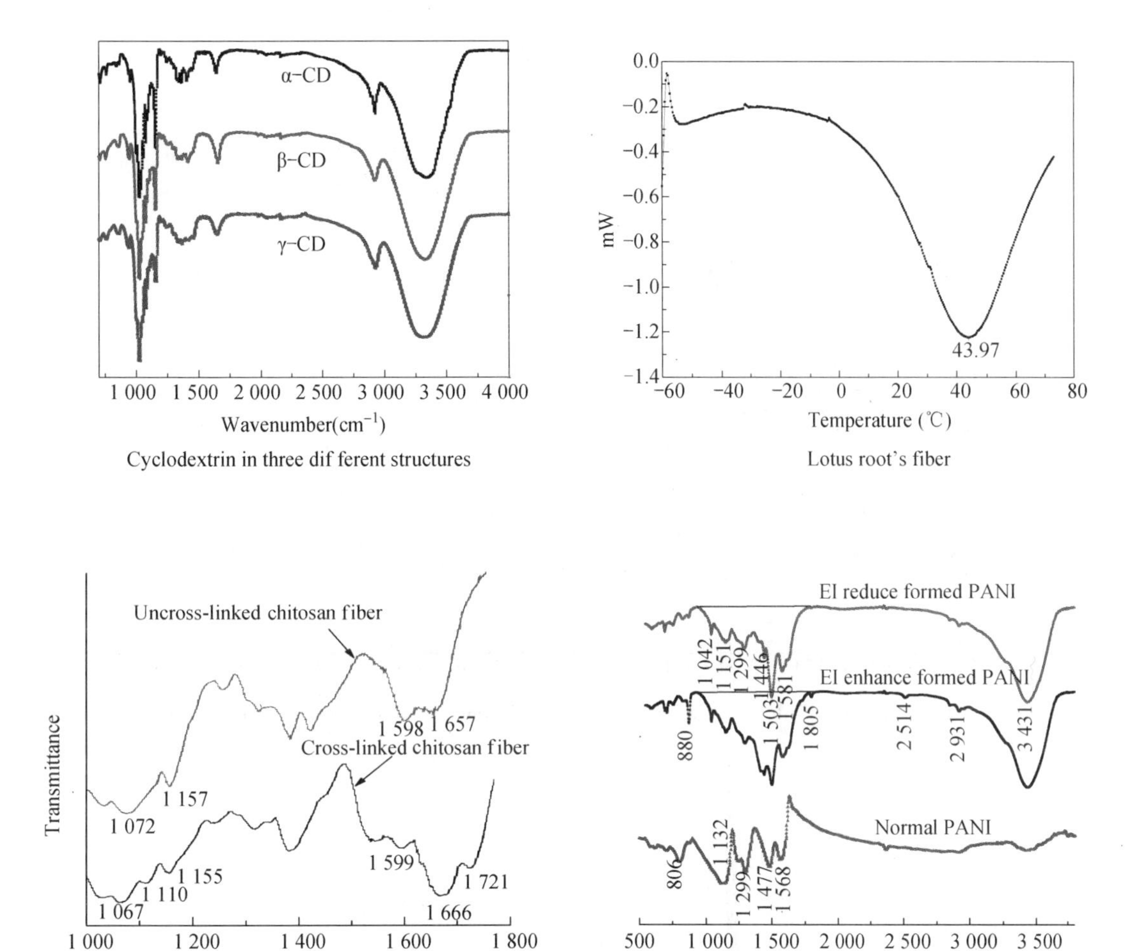

Figure 6-7 FTIR spectra of some materials

However, in some atoms (such as ^{1}H and ^{13}C) the nucleus does possess an overall spin.

NMR technique has two models in relation to solution and solid materials, respectively. Below presented some NMR spectra of lignin materials (Figure 6-8).

6.2.4 Ultraviolet visible spectroscopy

UV-vis spectroscopy is appropriate for characterization of aromatic or conjugated systems in fiber. The high sensitivity allows study of species at high dilution and is useful for intrinsically conducting polymers and lightemitting species. The absorption is characteristic of the π-π^* transition; fluorescence and phosphorescence lifetimes are influenced by the state of aggregation and the matrix structure in which the absorbing species is localized. Many pigments and dyes have characteristic absorption and emission spectra. Luminescence techniques are valuable for the study of degradation and aging in polymers.

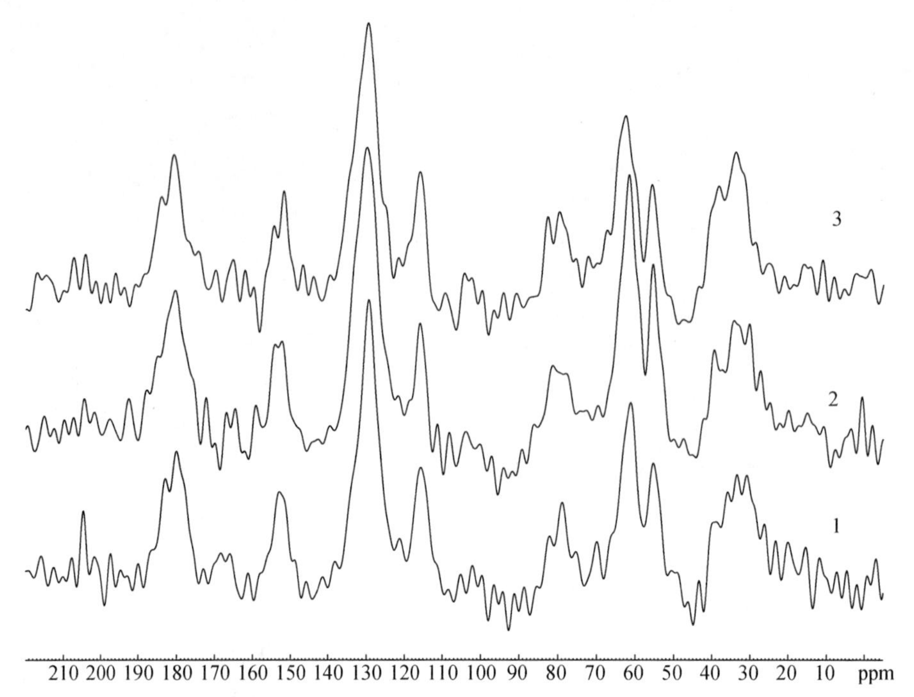

Lignin-based carbon films with different added lignin contents (solid state ^{13}C NMR spectra)

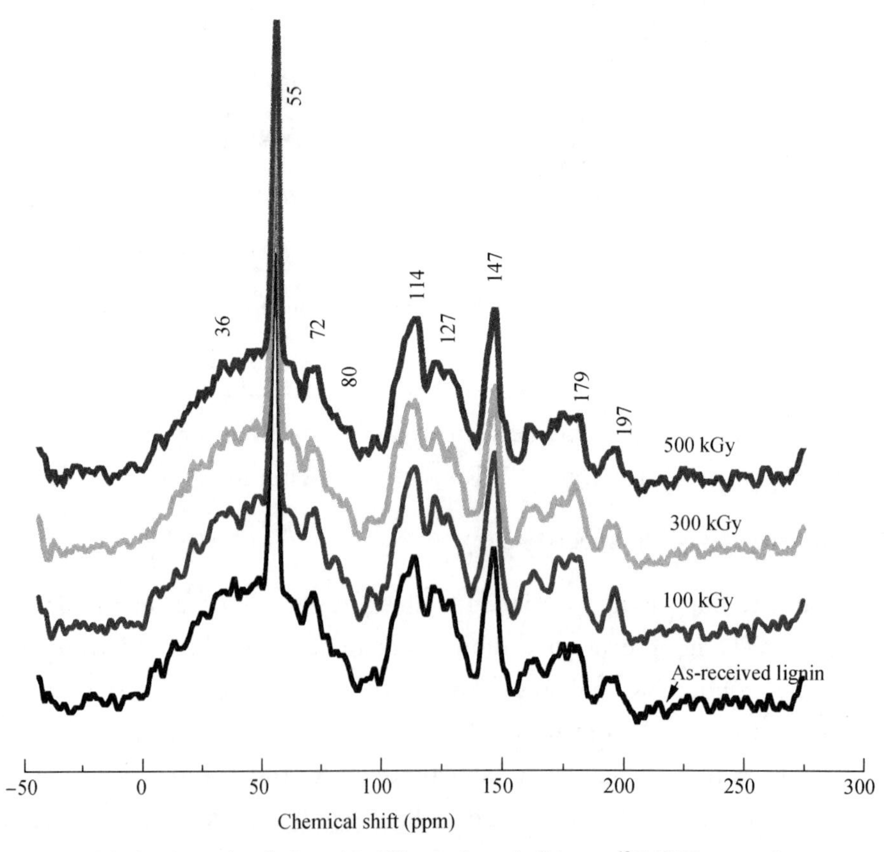

Lignin after γ-irradiation with different doses (solid state ^{13}C NMR spectra)

Figure 6-8 NMR spectra of various fibers

Rotation and local motion of groups in dilute solution on the order of $10^{-10} \sim 10^{-8}$s in amorphous polymers and orientated films may be studied by fluorescence depolarization. Doped polymer fiber excited with polarized light will exhibit fluorescence depolarization, which is characteristic of the motion in the system.

Luminescence may be suppressed by the presence of antioxidants, light stabilizers, and pigments, which are able to transfer energy form the excited state to other nonfluorescing electronic states. Quenching of excited states also occurs when either oxygen or moisture is present.

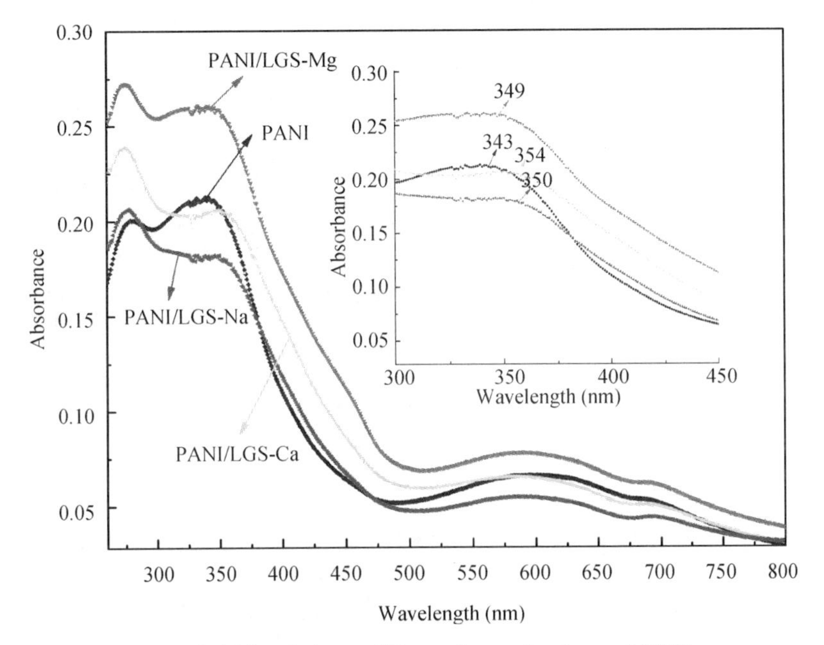

PANI/LGS in relation to different ions and referenced PANI

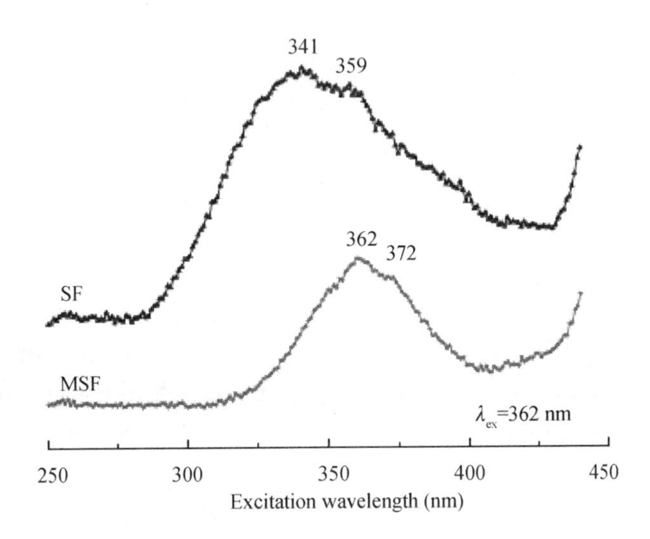

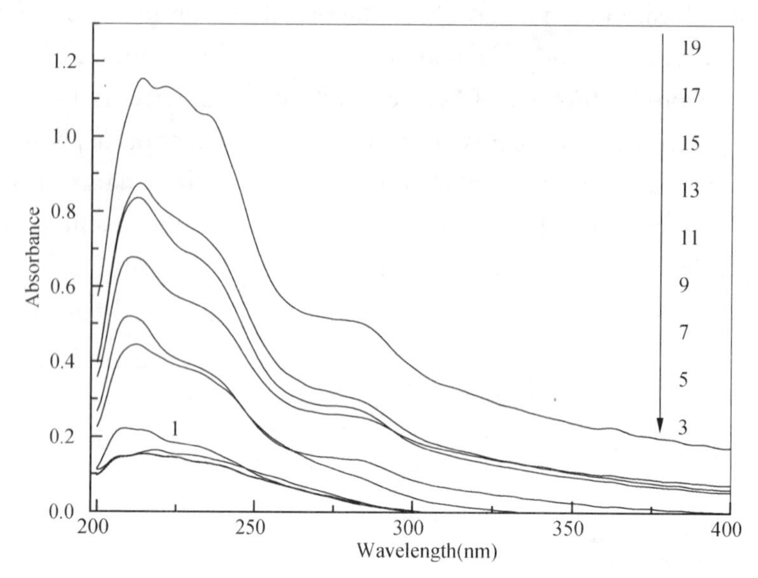

UV-spectra of self-assembled (CS/LGS)$_n$ multi-layer films

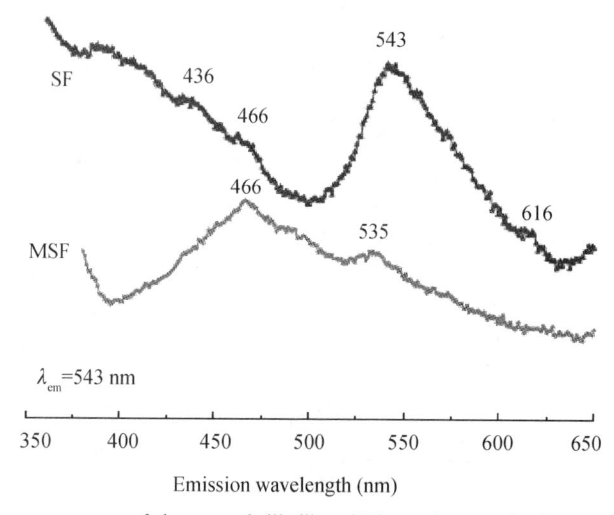

Fluorescence spectra of the normal silk fiber (SF), and magnetic silk fiber (MSF)

Figure 6-9 UV spectra of some materials including functional silk fibers

6. 2. 5 X-ray diffraction

X-ray diffraction, XRD, is a useful method for characterizing the structure of materials; especially the crystal structure and some chitosan-based XRD patterns were showned in Figure 6-10. Additionally, some fibers-based XRD patterns were showned in Figure 6-11.

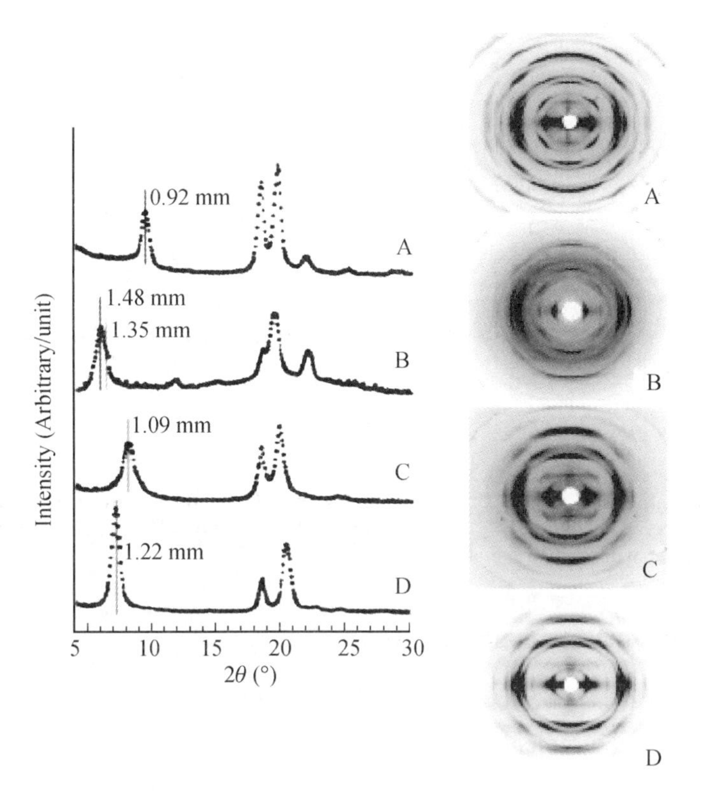

Figure 6-10 XRD patterns and equatorial profiles. (A) anhydrous *β*-chitin; (B) *β*-Chitin-acetic anhydride complex; (C) Acetylated *β*-chitin after 10 min of heating at 105 ℃; (D) Acetylated â-chitin after 2 h of heating at 105 ℃ (Yoshifuji et al. *Biomacromolecules*. 2006, 7, 2878-2881)

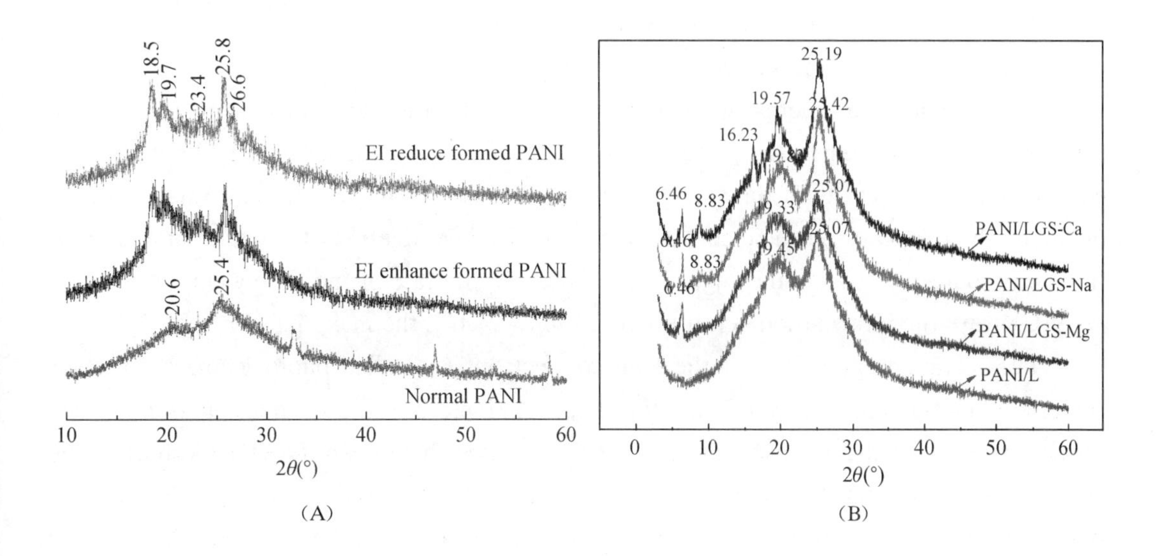

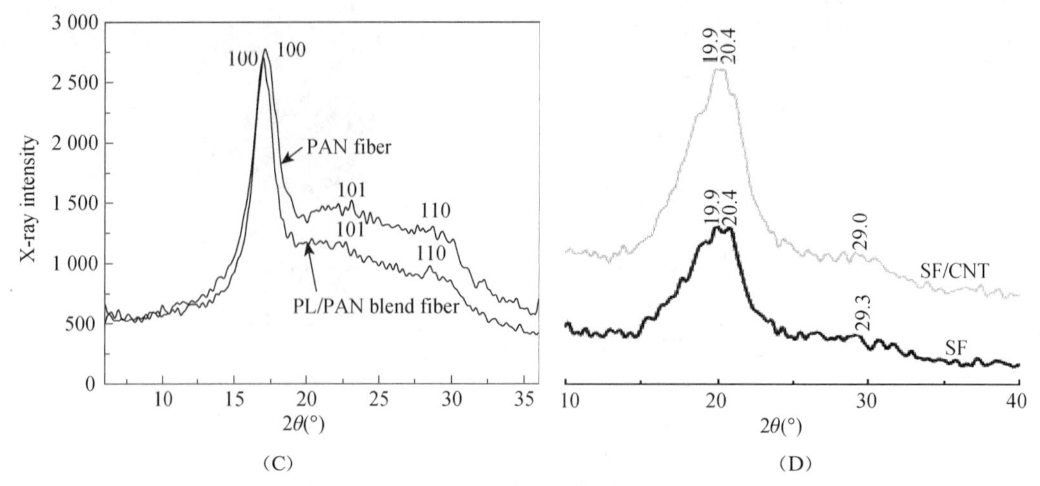

Figure 6-11 XRD patterns of various fibers. (A) PANI nanofibers formed by electro-synthesis; (B) PANI nanofibers guided by LGS; (C) PAN and PAN/persimmon leaves (PL) blending fiber; and (D) silk fiber with carbon nanotubes

6.2.6 X-ray photoelectron spectroscopy

X-ray photoelectron spectroscopy (XPS), which involves the measurement of the binding energies of electrons ejected by the ionizations of atoms with a monoenergetic beam of soft X-rays (Figure 6-12), has widely been used for the surface characterization of fiber and it-based materials. The XPS technique, developed by Siegbahn and co-workers, provides a unique tool for the investigation of a solid surface. Although X-rays penetrate deep into a sample, the XPS technique is very sensitive to surface constituents. Electrons emitted from the bulk of the material lose their energy with a high probability through collisions with electron orbitals of the bulk atoms. As a result, only atoms in a surface layer of very limited depth contribute to the intensity of the measured electron emission. Hence, the information depth of the method depends on the element which is under investigation, the quantum energy of the X-ray source ($h\hat{\imath}$), and the angle of incidence. In the case of the widely employed weak Al KR X-rays the information depth is *ca*. 10 nm which is the maximum for the C1s level.

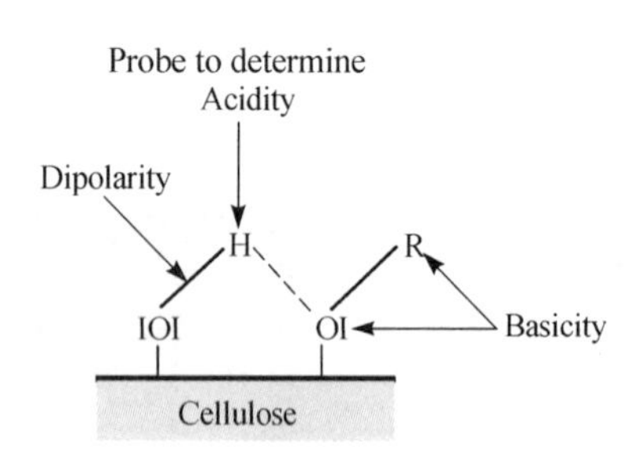

Figure 6-12 Typical acid/base interaction sites on a cellulose surface and their possible interactions with the different solvatochromic dyes

The kinetic energy (E_{kin}) of the emitted electrons (so-called photoelectrons) or their corresponding binding energy (E_{bin}) $h\hat{\imath}$ - E_{kin}) characterizes the elements present in the surface layer, and the intensity of the signal indicates their quantity. Furthermore, the energy of electrons emitted from a given element shell may be altered, depending on the type of chemical bond formed by the element. Therefore, on the basis of the possible chemical shift of the binding energy of a given element, the type of chemical bond present in the surface can be determined.

Figure 6-13 presented XPS survey spectrum of wood. In which, the C and O groups were strongly appeared indicating these elements are rich in wood than that of other appeared elements, e. g. N. In terms of the C1 peak, Figure 6 - 14 furthermore deconvoluted C1s spectrum of this wood. Observed there have three peaks located at 281～286 eV, 284～288 eV and 285～290 eV, corresponding to the C—C, C—O and C=O group, respectively.

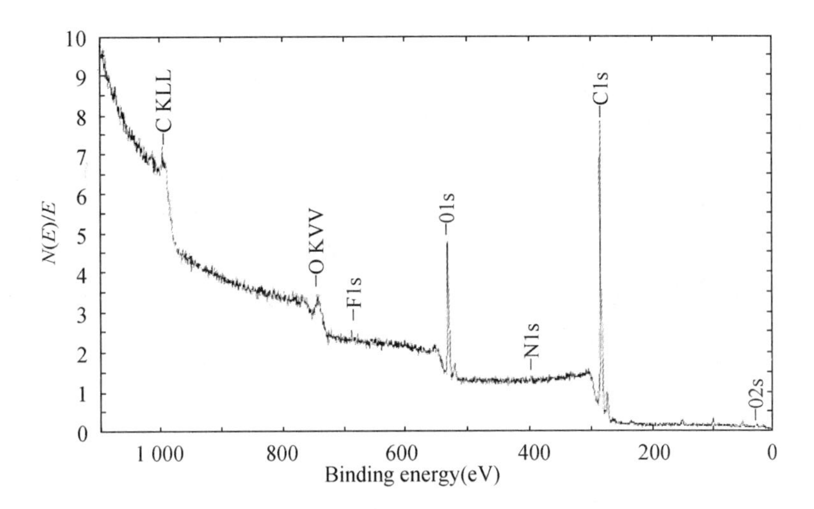

Figure 6-13 XPS survey spectrum of wood

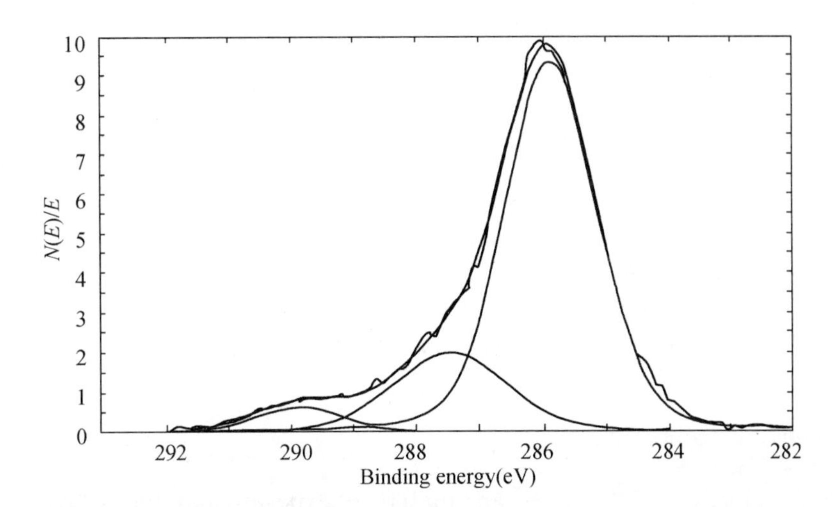

Figure 6-14 Deconvoluted C1s spectrum of wood presented three peaks located at 281～286 eV, 284～288 eV and 285～290 eV, corresponding to the C—C, C—O and C=O group, respectively

6.3 Properties characterization

6.3.1 Mechanical properties

The measurement of mechanical properties is concerned with load-deformation or stress-strain relationships. Forces may be applied as tension, shear, torsion, and compression and bending. Stress is the force divided by the cross-sectional area of the sample. Strain is the change in a physical dimension of the sample divided by the original dimension. The ratio of the stress to strain is referred to the modulus. Stress may be applied continuously or periodically at varying rates for different tests. The characteristic stress-strain curve, stress relaxation, or impact behavior is very important in determining the applications and limitations of a polymer fiber.

During these measurements, the gauge length and crosshead speed were chosen, e.g. at 500 mm and 7.5 mm/min, respectively, in accordance with ASTM D3822-1990. Since the fiber quality to be varied with spinning process, each given value would be averaged by several independent runs using different samples.

Figure 6-15 showed some mechanical properties of materials.

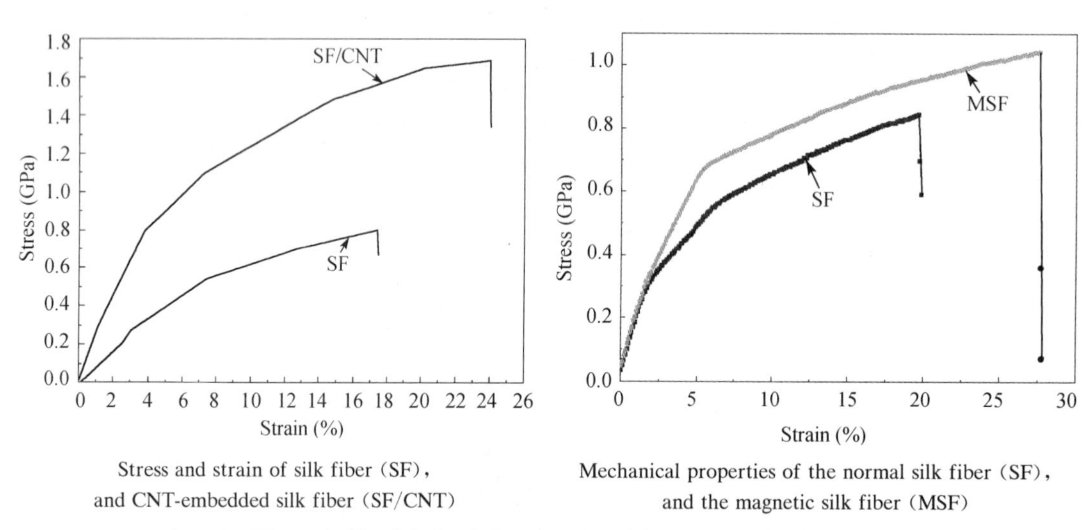

Stress and strain of silk fiber (SF), and CNT-embedded silk fiber (SF/CNT)

Mechanical properties of the normal silk fiber (SF), and the magnetic silk fiber (MSF)

Figure 6-15 Mechanical properties of functional silk fibers

6.3.2 Thermal properties

Thermal analysis is defined as an analytical experimental technique and can be applied to investigate the physical properties of fiber how be influenced by temperature or time under controlled conditions. The main thermal techniques are referred to thermogravimetry, TG, differential thermal analysis, DTA, differential scanning calorimetry, DSC, thermomechanometry, TMA, and dynamic mechanical analysis,

DMA, respectively.

Recently, simultaneous measurements combining various techniques are widely used and this extended to thermal analysis to be the TG-DTA, TG-FTIR.

Figures 6-16, 17 and 18 presented the typical TG, DSC and DMA curves of some materials including some fibers. In these methods, the glass translation temperature, T_g, could be determined only by DSC and DMA, and of the latter the presented α peak corresponds to the T_g.

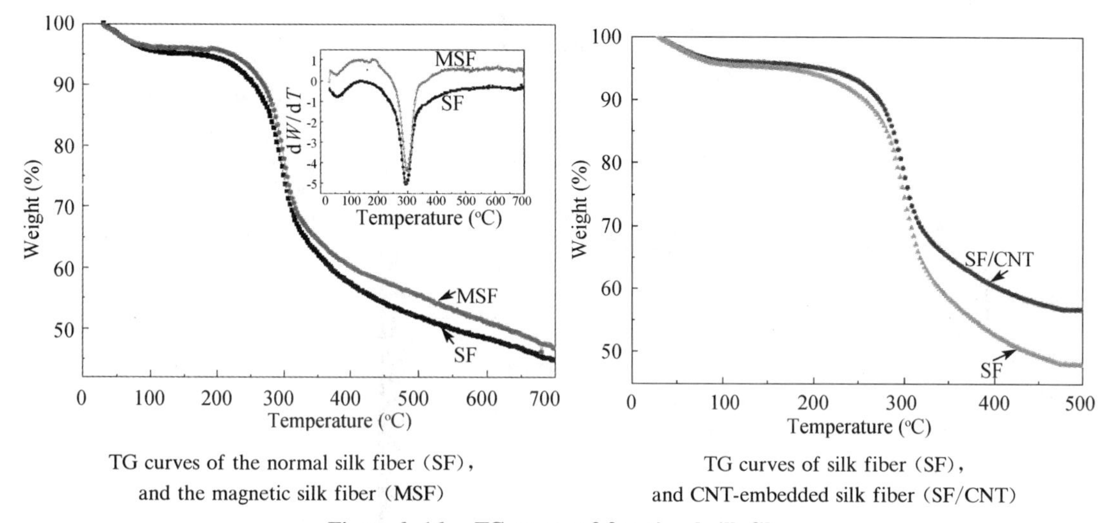

TG curves of the normal silk fiber (SF), and the magnetic silk fiber (MSF)

TG curves of silk fiber (SF), and CNT-embedded silk fiber (SF/CNT)

Figure 6-16 TG curves of functional silk fiber

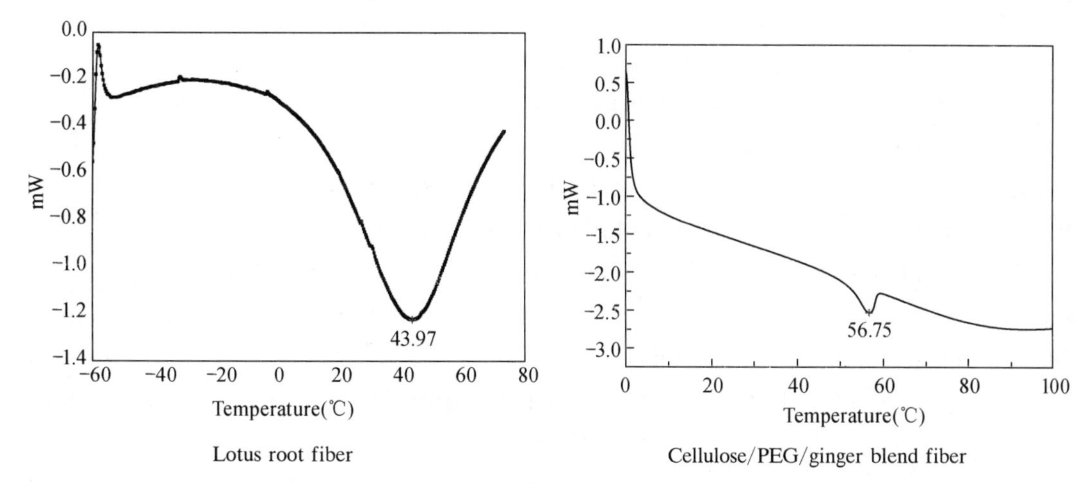

Lotus root fiber

Cellulose/PEG/ginger blend fiber

Figure 6-17 DSC curves of some fibers

6.3.3 Electric and magnetic properties

Although polymer fibers have been traditionally used as insulators, in recent years the ability to achieve intrinsic conductivity fiber has been demonstrated and is associated with systems in which the backbone has developed an extended delocalized electronic

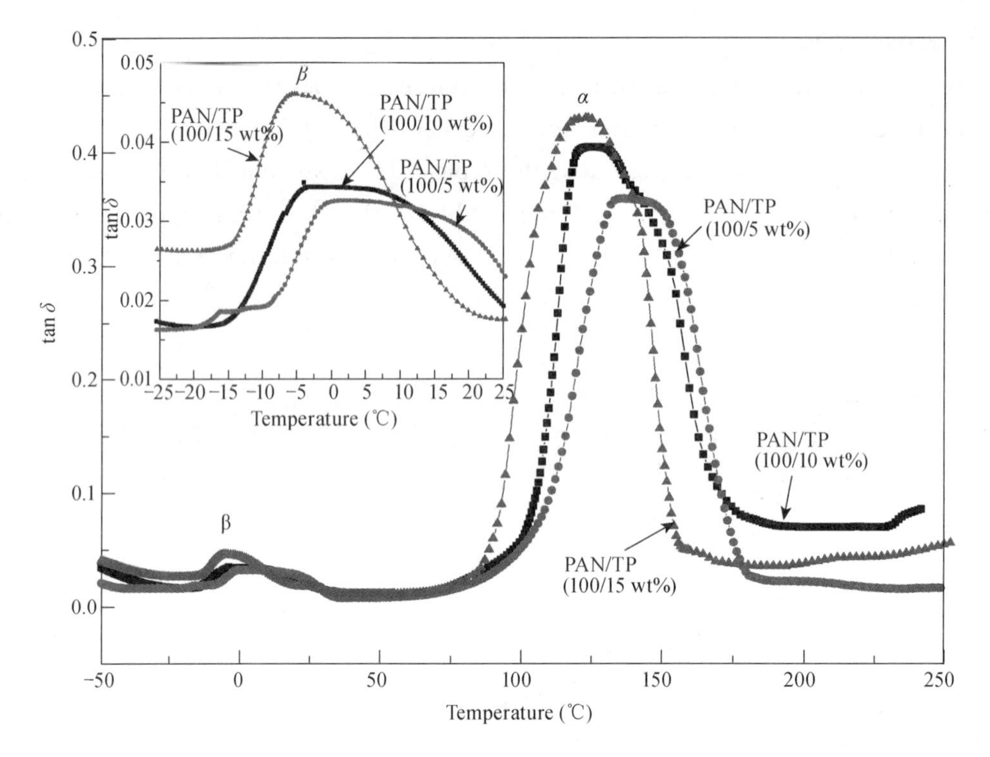

Figure 6-18　DMA curves of PAN/TP blends

structure. A range of tests exists to help measure such features as the electric breakdown resistance. Other specific types of property observed in polymers include semiconductivity, photoconductivity, piezoelectricity, pyroelectricity, and static electrical charging (triboelectricity). Each of these features is a direct consequence of a combination of chemical and specific morphological feature in the polymer or composite material.

The electrical conductivity of fiber can be measured using SDY-4 Four-Point Probe Meter (Four Dimensions, Inc. USA) at 25 ℃. The sample can be prepared by subjecting as the powder to a pressure of 30 MPa, and the reproducibility can be checked by measuring the resistance of each sample in at least three times then averaged.

The magnetic property can be measured using 2 900 MicroMag Alternating Gradient Magnetometer at room temperature, 25 ℃. The apparatus was calibrated using a nickel foil with a saturation magnetization of 2. 144 emu.

Table 6-1 presented the electric and magnetic properties of functional silk fibers, and the latter was measured according to Figure 6-19.

Table 6-1　The electric and magnetic properties of functional silk fibers (SF—silk fiber, CNT—carbon nanotubes, MSF—magnetic SF)

Samples	Modules (GPa)	Stress at break (GPa)	Strain at break (%)	Electric resistance (Ω)
SF	0. 14±0. 05	0. 80±0. 3	17. 5±0. 3	211±10
SF/CNT	0. 38±0. 05	1. 69±0. 3	24. 0±0. 3	171±10

(Continued)

Samples	Saturated magnetic intensity (emu/g)	Retentivity (emu/g)	Coercivity (Oe)
SF	0	0	0
MSF	0. 3	0. 02	84
MSF	1. 5	0. 07	25
MCF	3. 7	0. 53	141

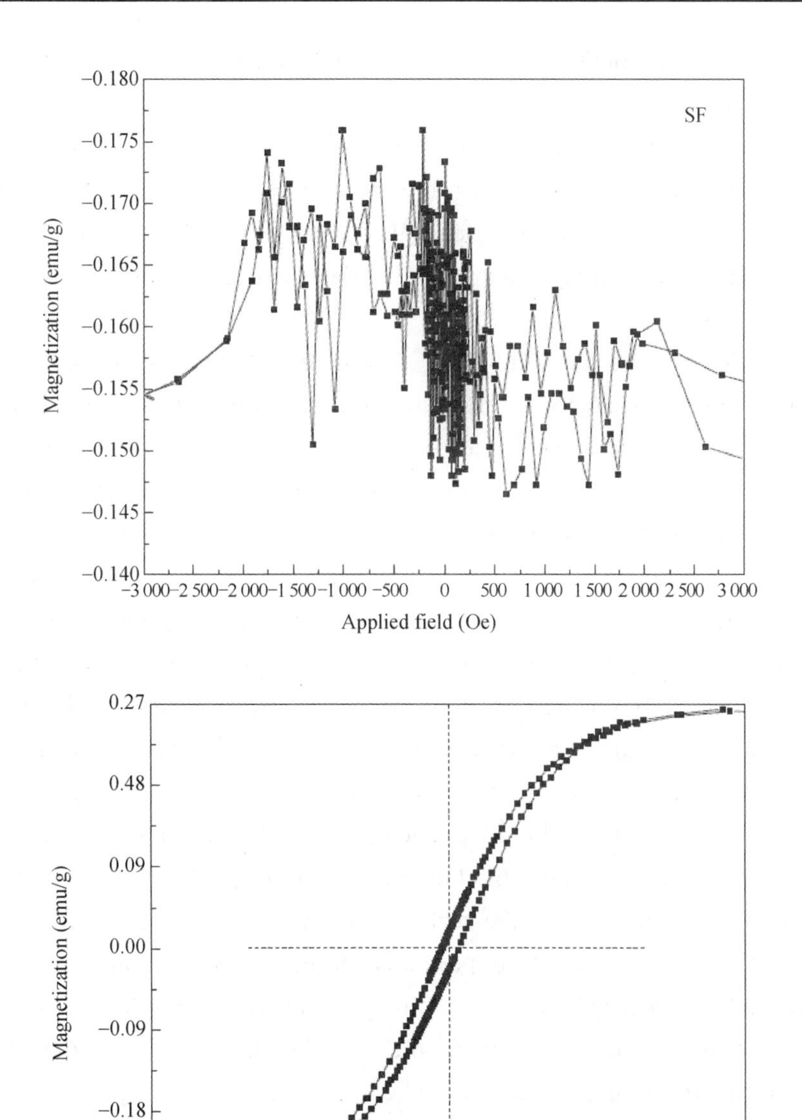

Figure 6-19 Magnetic properties of magnetic silk fiber (MSF), and a comparison with normal silk fiber (SF)

6.4 Surface behavior characterization

6.4.1 Wetting

Although a liquid will always try to form a minimum-surface-area shape, if no other forces are involved, it can also interact with other macroscopic objects, to reduce its surface tension via molecular bonding to another material, such as a suitable solid. Indeed, it may be energetically favorable for the liquid to interact and "wet" another material.

The wetting properties of a liquid on a particular solid are very important in many everyday activities and are determined solely by surface properties. One important and common example is that of water on clean glass. Water wets clean glass because of the favorable hydrogen bond interaction between the surface silanol groups on glass and adjacent water molecules.

This dramatic macroscopic difference in wetting behavior is caused by only a thin molecular layer on the surface of glass and clearly demonstrates the importance of surface properties. The same type of effect occurs every day, when dirty fingers coat grease onto a drinking glass! Surface treatments offer a remarkably efficient method for the control of macroscopic properties of materials. When insecticides are sprayed onto plant leaves, it is vital that the liquid wet and spread over the surface. Another important example is the froth flotation technique, used by industry to separate about a billion tons of ore each year. Whether valuable mineral particles will attach to rising bubbles and be "collected" in the flotation process, is determined entirely by the surface properties or surface chemistry of the mineral particle, and this can be controlled by the use of low levels of "surface-active" materials, which will selectively adsorb and change the surface properties of the mineral particles. Very large quantities of minerals are separated simply by the adjustment of their surface properties.

It is easy to demonstrate that the surface energy of a liquid actually gives rise to a "surface tension" or force acting to oppose any increase in surface area. Thus, we have to "blow" to create a soap bubble by stretching a soap film. A spherical soap bubble is formed in response to the tension in the bubble surface. The soap film shows interference colors at the upper surface, where the film is starting to thin, under the action of gravity, to thicknesses of the order of the wavelength of light.

6.4.1.1 Measuring the surface tension of liquid

The measurement of the surface tension, γ_L, of liquid has a lot of methods and herein we introduced only two often used methods. The first is the capillary rise method and the principle was described by Figure 6-20.

In this method the height to which the liquid rises, in the capillary, above the free liquid surface is measured. This situation is illustrated in Figure 6 - 20. Using the Laplace equation the pressure difference between points A and B is given simply by $\Delta P = 2\gamma_L/r$, if we assume that the meniscus is hemispherical and of radius r. However, this will be accurate only if the liquid wets the walls of the glass tube. If the liquid has a finite contact angle θ with the glass, then from simple geometry (again assuming the meniscus is spherical).

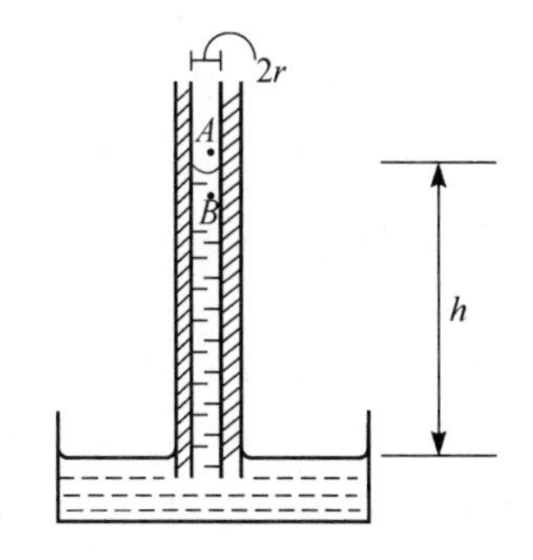

Figure 6 - 20 Schematic diagram of the rise of a liquid that wets the inside walls of a capillary tube

$$\Delta P = 2\gamma_L \cos\theta / r \tag{6-1}$$

Note that if $\theta > 90^\circ$ (e. g. mercury on glass), the liquid will actually fall below the reservoir level and the meniscus will be curved in the opposite direction.

The pressure difference between points A and B must be equal to the hydrostatic pressure difference $h\rho g$ (where ρ is the density of the liquid and the density of air is ignored). Thus, we obtain the result as:

$$\gamma_L = rh\rho g / 2\cos\theta \tag{6-2}$$

Another common method used to measure the surface tension of liquids is called the "Wilhelmy plate". These methods use the force (or tension) associated with a meniscus surface to measure the surface energy rather than using the Laplace pressure equation. (Note that in real cases both factors usually arise but often only one is needed to obtain a value for γ_L.) The Wilhelmy plate is illustrated in Figure 6-21.

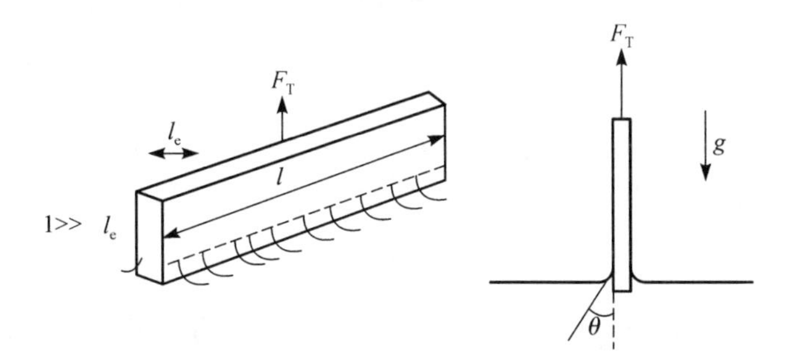

Figure 6-21 Diagram of the Wilhelmy plate method for measuring the surface tension of liquids

The total force F_T (measured using a balance) is given by

$$F_T = F_W + 2l\gamma_L \cos\theta \tag{6-3}$$

Where F_W is the dry weight of the plate. (Note that the base of the plate is at the same level as the liquid thereby removing any buoyancy forces). The plates are normally made of thin platinum which can be easily cleaned in a flame and for which l_c can be ignored. Again, this method has the problem that θ must be known if it is greater than zero. In the related du Noüy ring method, the plate is replaced by an open metal wire ring. At the end of this chapter, a laboratory class is used to demonstrate yet another method, which does not require knowledge of the contact angle and involves withdrawal of a solid cylinder attached to a liquid surface.

6.4.1.2 Estimation of the surface free energy of solid materials

There are basically two ways by which we can attempt to obtain the surface energy of solids: the first is the measurement of the cohesion of the solid, and the second is to study the wetting behavior of a range of liquids with different surface tensions on one unknown solid surface. Yet neither methods is straightforward and the results are not as clear as those obtained for liquids.

The cohesive energy per unit area, W_c, is equal to the work required to separate a solid in the ideal process illustrated in Figure 6-22. In this ideal process the work of cohesion, W_c, must be equal to twice the surface energy of the solid, γ_S. Although this appears simple as a thought experiment, in practice it is difficult. For example, we might measure the critical force (F_c) required to separate the material but then we need a theory to relate this to the total work done. The molecules near the surface of the freshly cleaved solid will rearrange after measuring F_c. Also, the new area will not usually be smooth and hence the true area is much larger than the geometric area.

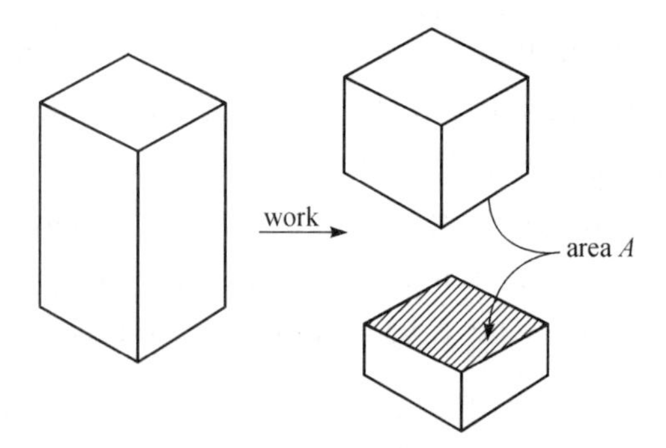

Figure 6-22 Ideal experiment to measure the work required to create new area and hence find the surface energy of a solid

The adhesive energy per unit area W_a between two different solids is given by Figure 6-23 and Eq.(6-4):

$$W_a = \gamma_A + \gamma_B - \gamma_{AB} \tag{6-4}$$

Where γ_A and γ_B are the surface energies of the solids and γ_{AB} is the interfacial energy of the two solids in contact ($\gamma_{AA} = 0$).

Again the adhesive energy is a difficult property to measure. It is also very hard to find the actual contact area between two different materials since this is almost always much less than the geometric area. That this is the case is the reason why simply pressing two solids together does not produce adhesion (except for molecularly smooth crystals like mica) and a "glue" must be used to dramatically increase the contact area. The main function of a glue is to facilitate intimate molecular contact between two solids, so that strong short-range van der Waals forces can hold the materials together.

The second approach to obtaining the surface energies of solids involves the study of wetting and non-wetting liquids on a smooth, clean solid substrate. Let us examine the situation for a non-wetting liquid (where $\theta > 0^\circ$), which will form a sessile drop on the surface of a solid (Figure 6-23). Using an optical microscope, it is possible to observe and measure a finite contact angle (θ) as the liquid interface approaches the three-phase-contact perimeter of the drop.

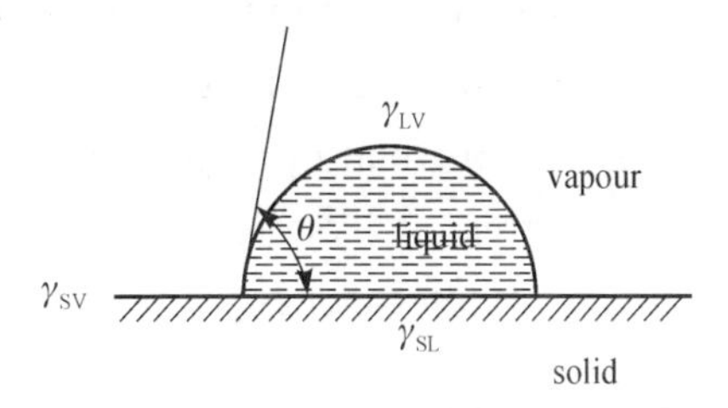

Figure 6-23 Diagram of the wetting of solid by a probe liquid

The total interfacial energy change must be given by the sum:

$$\gamma_S = \gamma_L + \gamma_L - \gamma_L \cos\theta \tag{6-5}$$

Since one can measure the liquid surface energy, γ_L, the value of γ_S can be obtained by measuring only the θ.

Clearly the surface energy of a solid is closely related to its cohesive strength. The higher the surface energy, the higher is its cohesion. This has some obvious and very important ramifications. For example, the strength of a covalently bonded solid, such as a glass or metal, must always be greatest in a high vacuum, where creation of new surface must require the greatest work. The strength of the same material in water vapor or immersed in liquid water will be much reduced, often by at least an order of magnitude. This is because the freshly formed solid surface must initially be composed of high-energy atoms and molecules produced by the cleavage of many chemical bonds. These new high-energy surfaces will rapidly adsorb and react with any impingent gas molecules. Many construction materials under strain will therefore behave differently, depending on the environment.

Although the contact angle is a very useful indicator of the energy of a surface, it is also affected by a substrate's surface roughness and chemical micro-heterogeneity. This can be well illustrated by comparing ideal, calculated contact angle values with measured ones. For example, one can easily calculate the expected water contact angle on a liquid or solid pure hydrocarbon surface by using the surface tension of water and hexadecane, at about 27.5 mN/m, and the interfacial tension between the oil and water, e.g. at about 53.8 mN/m. Use of these values in the Young equation gives an expected angle of 111°, in close agreement with the values observed for paraffin wax. However, on many real surfaces the observed angle is hysteretic, giving quite different values depending on whether the liquid droplet size is increasing (giving the advancing angle θ_A) or decreasing (giving the receding angle θ_R). The angles can differ by as much as 60° and there is some controversy as to which angle should be used (for example in the Young equation) or even if the average value should be taken. Both θ_A and θ_R must be measured carefully, whilst the three-phase line is stationary but just on the point of moving, either forwards or back. In general, both angles and the differences between the two give indirect information about the state of the surface, and both should always be reported. The degree of hysteresis observed is a measure of both surface roughness and surface chemical heterogeneity.

6.4.2 Coating

Coating is often used to modify solid surface by deposition of some functional groups-based materials. Observed the hydrophobic surface can be turned into hydrophilic or on the contrary.

Recommending reading

[1] Q. Shen, J. B. Rosenholm. *Characterization of the Wettability of Wood Resin by Contact Angle Measurement and FT-Raman Spectroscopy*. In: *Advances in Lignocellulosics Characterization*. Chapter 10, Ed. D. S. Argropoulos, TAPPI Press, 1999.

[2] Q. Shen. *Surface Properties of Cellulose and Cellulose Derivatives*. In: *Model Cellulosic Surfaces*. Chap. 12, Ed. Oxford University Press, 2009.

Problems

1. How to measure the surface free energy of a solid material?

Chapter 7

Various Cases on Fiber Formation

7.1 Case on directly formation of super strength silk fiber from silkworm

Silk fiber, SF, is the example of man-made fiber and has leaded a lot of development in fiber world. Bombyx mori SF is a semicrystalline biopolymer with 80%~85% of glycine, alanine and serine. SF has for a long time been applied in textile, biotechnological and biomedical fields due to its high strength, rupture elongation, environment stability and biocompatible properties. In order to obtain SF with enhanced mechanical or functional properties, a lot of work has been tried as can be seen elsewhere which includes the chemical modifications, adjustment of the harvesting parameters and reconstruction of SF via the artificial spinning by adding some nano- or functional materials to mix with silk protein. The use of carbon nanotube, CNT, to reinforce SF has been tried in an artificial spinning process and the presented regenerated SF, RSF, has been found to have enhanced mechanical and electrical properties. In addition, it is also noted that Shao and Vollrath have reported a special case to obtain reinforced pristine SF by fixing the head of silkworm to forbid its randommotion during the fiber spinning process and meanwhile to increase the spinning speed by using an extra winder. According to these researchers' reported values, the strength enhancement is limited for SF because the stress increase is accompanied with the strain loss or on the contrary.

In this work we reported a simple method to obtain in vivo reinforced SF directly from silkworm. Experimentally, we reared twelve silkworms at the laboratory, and fed them with two kinds of mulberry leaves, MLs. One is the normal ML as obtained from the tree and another is the ML pretreated by spreading lignosulfonate, LGS, to modify CNT on the ML surface to form ML/CNT.

In this case, the used CNTs are multiwalls type with the purity above 90% purchased from Chengdu Institute of Organic Chemistry, Chinese Academy of Sciences. As known, these CNTs were formed by the chemical vapor deposition and their lengths are ranged from several hundred nanometers to several micrometers with an average outer diameter about 10~30 nm. The used LGS is a calcium-based sample with an average molecular weight of 100 000 provided by Jiangmen Sugar Cane Chemical Factory, Guangdong of China. According to the producer, this LGS is composed of phenylpropane segments and sulfuric acid groups, and its lignin component is more than 55%, its deoxidized sugar is less than 12%, its water-insoluble components are smaller than 1.5%, and its moisture is at about 9%. The pH of this LGS is known to be in the range of 5～6. This LGS was directly used as received without further purification.

The modification of CNTs by LGS was started by mixing 0.5 g CNTs with 5 g LGS in an agate mortar for 3 h by hand. During this process, a small amount of water was added to avoid agglomeration. The obtained brownish slurry was added to the 200 mL buffer and was subsequently sonicated using an ultrasonic tip to stimulate the micelle formation. After that, the mixed solution was again sonicated for about 30 min and centrifuged at 15 000 r/min to remove the residual solid mass. The obtained LGS treated CNTs were washed three times by 300 mL distilled water for each one to remove the excess LGS. The prepared LGS treated CNTs were finally oven dried with measured moisture at about 10% then applied to spread on the ML surface to feed silkworms.

The use of LGS to modify CNT is based on two reasons, the first is that the LGS is also a natural biomaterial obtained from plant closing the MLs that may fit the worms, and the second is that the LGS is usually associated with some metal ions thus considered to help CNT mix well with silk protein.

The obtained cocoons were immersed in warm water (about 80～90 ℃) with a pH of about 9～10 for about 1～3 h then the SF was hand winded on a glass bottle surface. The collected SF was prepared in a length of about 2～3 m then oven dried at 90 ℃ for 24 h to keep the moisture constant in a range of 3%～5%.

The SEM image provided a possibility to compare the morphology of both the SF and SF/CNT (Figure 7-1). According to Figure 7-1, the cross-section of normal SF is in triangle (Figure 7-1, top) and is in elliptic shape for SF/CNT (Figure 7-1, bottom). These differences suggested that the CNTs are indeed embedded in SF and the CNT embedding caused silk protein reconstruction. According to Figure 7-1, the fiber thickness seems to be unaffected by CNT-embedding because these two SFs showed a similarly thickness at about 10～20 μm. Since a fine needle, e.g. its diameter at about 100 nm, was obviously inserted in the cross-section of SF/CNT (Figure 7-1, bottom), this directly indicated that the CNT indeed embedded in SF. In fact, in terms of the known original diameter of CNT, e.g. about 30 nm, it is known that the LGS coating and silk protein sheathing caused the total surface thickness of CNT increased about 35 nm. This is reasonable because the used LGS is water soluble and is capable of coating the CNT surface.

In terms of Figure 7-1, the SF is actually a duplet of two individual fibrous with their own silk coating (sericin) and an inner core (fibroin) and structured by thousands of parallel fibrils (100～400 nm) in good agreement with literature. A comparison of the

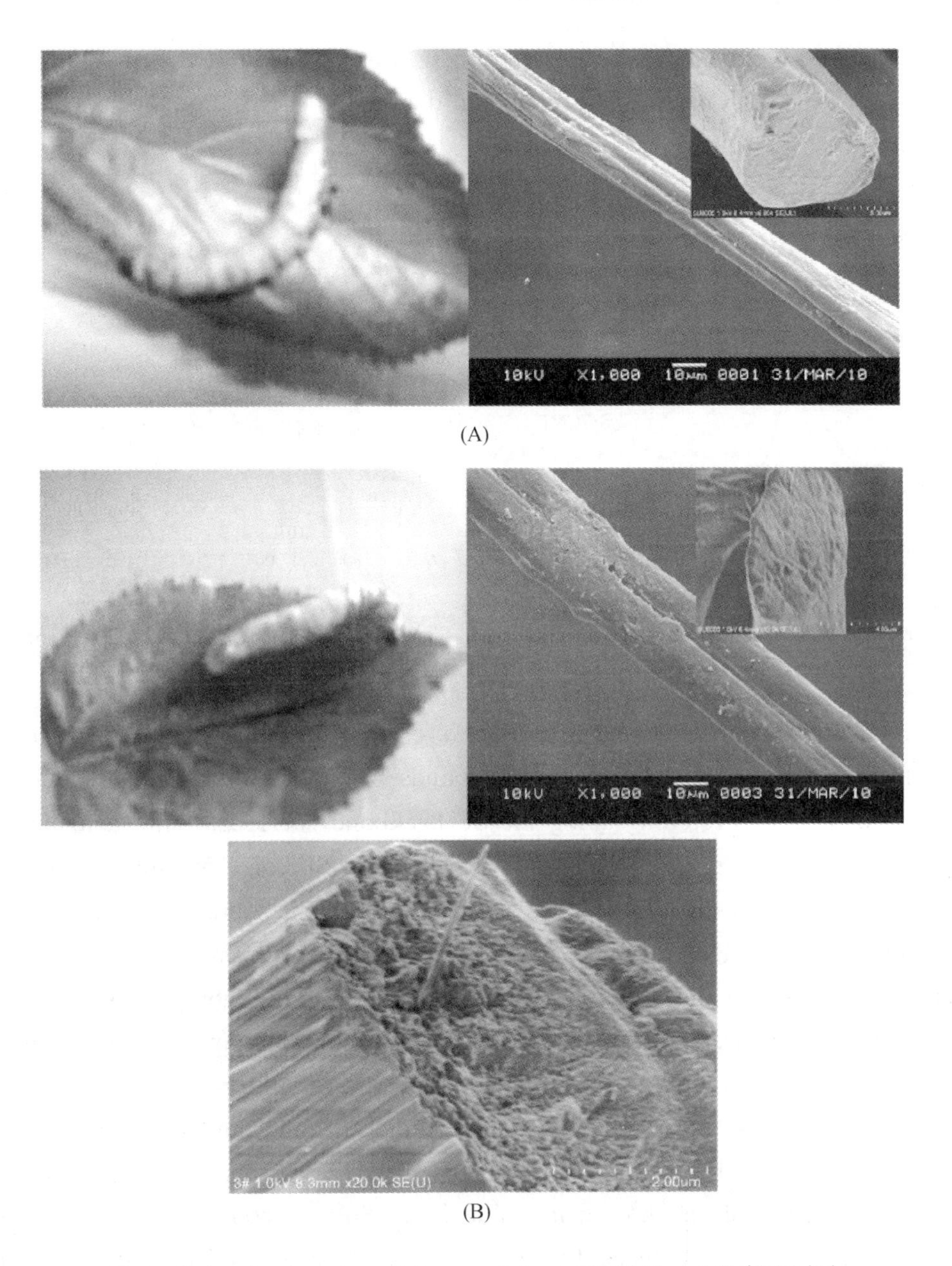

Figure 7-1 A comparison of the morphology of SF (A) and SF/CNT (B)

fiber surface found that the normal SF has a smooth surface while the SF/CNT showed a lot of sort of debris on the surface to suggest that the presence of CNT in silk protein caused some influences on silkworm spinning. This is also possible because the CNTs might be partly free to move without sheared by silk protein and thus be separated by the spinning process-associated extrusion force due to the random motion of worm head.

The XRD patterns of SF and SF/CMT were presented and compared in Figure 7-2. Though these three SFs showed three main 2θ peaks corresponding to the silk-I and -II structures, respectively, some differences are visible because the second peak was found to locate at 20.8° for SF and at 20.4° for SF/CNT, and the third peak was located at 29.3°

for SF and at 29. 0° for SF/CNT. Clearly, these peaks shifted downward or upward both due to the presence of CNT in silk protein and this caused the changes on silk-I structure.

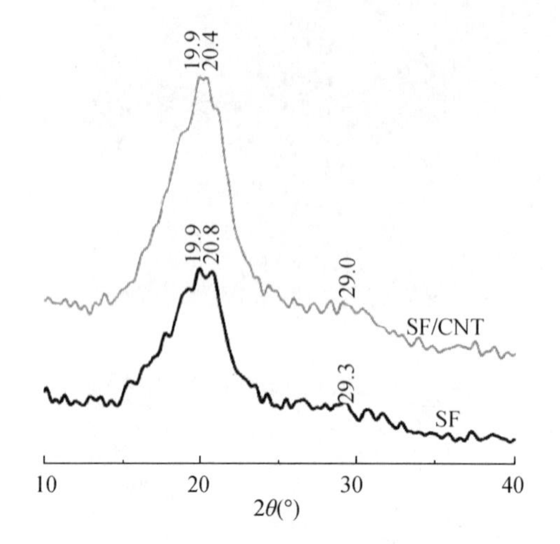

Figure 7-2 A comparison of the XRD patterns of SF and SF/CNT

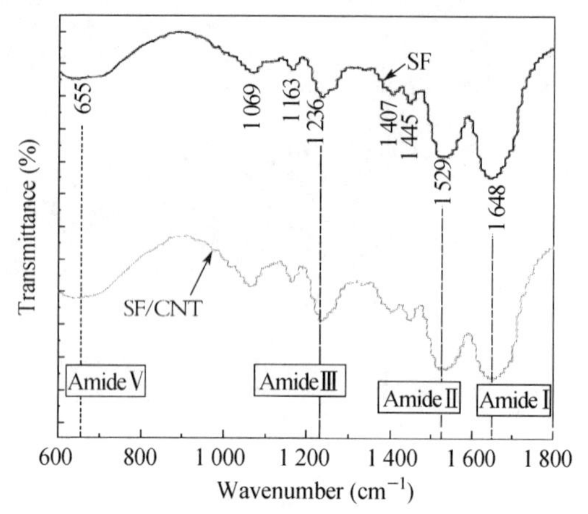

Figure 7-3 A comparison of the FTIR spectra of SF and SF/CNT

The FTIR spectra of SF and SF/CNT were presented in Figure 7-3. In terms of literature, the band at 1 648 cm^{-1} corresponds to the amide-I structure due to the indication of C=O stretching along the SF backbone, at 1 529 cm^{-1} assigned to the amide-II structure due to the N—H deformation in the SF corresponding to the β-sheet, at 1 234 cm^{-1} assigned to the amide-III structure due to the vibrations of O—C—O and N—H corresponding also to the β-sheet, and at 655 cm^{-1} assigned to the amide-V structure in relation to the β-sheet. The noted two SFs both showed a peak located at about 1 407 cm^{-1} related to the vibrations of CH groups to imply that the embedding CNTs in SF does not greatly change the inter-structure of silk protein.

Taking the peak intensity of amide-II/amide-I and amide-III/amide-I as two structure parameters, the structure changes on SF/CNT was quantitatively evaluated to compare the normal SF as showed in Table 7-1. Since a comparison of two samples found that the SF/CNT reduced the amide-II structure about 5% and increased the amide-III structure about 10% as that of the SF, this clearly indicated that the SF/CNT was reconstructed due to CNT-embedding.

Table 7-1 Structure parameters of SF and SF/CNT on the basis of FTIR spectra

Samples	Amide-II/Amide-I	Amide-III/Amide-I
SF	0. 75 ± 0. 05	0. 45 ± 0. 05
SF/CNT	0. 70 ± 0. 05	0. 60 ± 0. 05

The thermal properties of two SFs were compared by TGA by increasing the temperature from 25 ℃ to 500 ℃ in a nitrogen atmosphere. The recorded TG curves were showed in Figure 7-4, where both SFs have about 5% moisture which was fast removed during the initial temperature increase stage, e. g. from 25 ℃ to 100 ℃. According to Figure 7-4, the normal SF has a degradation temperature at about 170 ℃, and this temperature would be enhanced to about 217 ℃ for SF/CNT. This finding has not only again proven the presence of CNT in SF but also importantly indicated that the CNT-embedding in SF can enhance the thermal stability of SF.

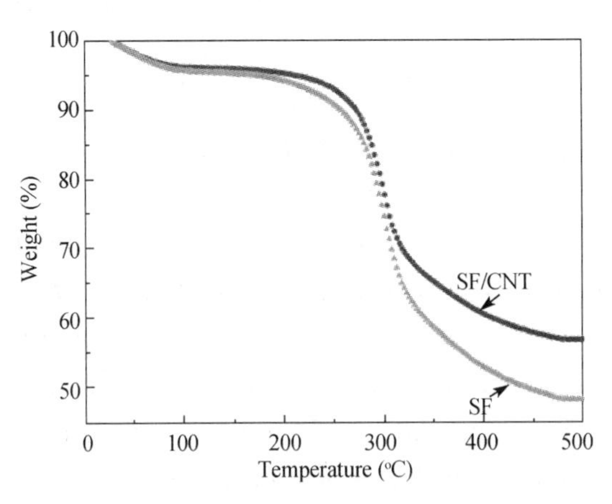

Figure 7-4 A comparison of the TG curves of SF and SF/CNT

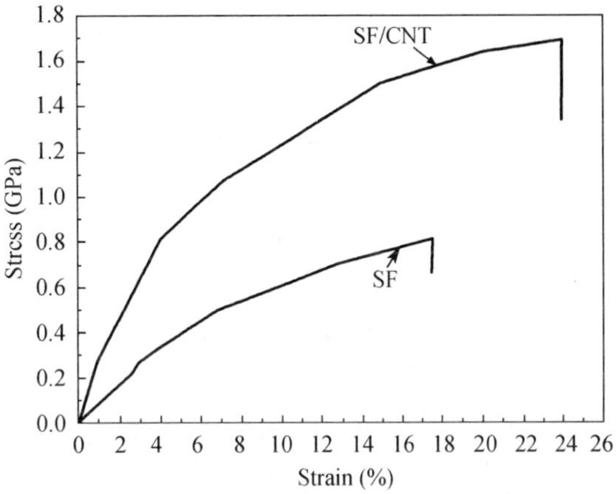

Figure 7-5 A comparison of the mechanical properties of SF and SF/CNT

The mechanical properties of these two SFs were compared in Figure 7-5 where the maximum values were also summarized in Table 7-2. It is of interest that the SF/CNT presented a higher stress at 1. 69 GPa and a higher strain at 24. 0% because the former is about 110% greater and the latter is about 37% greater than that of normal SF, respectively. Since RSF has presented a strain at about 20% and the super SF prepared by Shao and Vollrath has presented a stress at about 1. 0GPa corresponding to a strain at about 20%, it is interesting that Table 7-2 presented values indicating that in this case we obtained SF/CNT having high strength. In fact, this SF/CNT presented mechanical properties that are comparable with those of the spider fiber.

The measured electric resistance of these two SFs were summarized and compared in Table 7-2. Since the SF/CNT was expected to present enhanced conductivity when compared with SF, this reasonably implied that the CNTs are aligned to embed in SF. According to Table 7-2, this SF/CNT has some new functions and is obviously a novel SF capable of fitting the request of some new application cases.

Table 7-2 A comparison of the mechanical and electric properties of SF and SF/CNT

Samples	Modules (GPa)	Stress at break (GPa)	Strain at break (%)	Electric resistance (Ω)
SF	0. 14±0. 05	0. 80±0. 3	17. 5±0. 3	211±10
SF/CNT	0. 38±0. 05	1. 69±0. 3	24. 0±0. 3	171±10

In this work, a simple process was developed for obtaining high strength SF directly from silkworm. By feeding the worms with CNT, experiments have proven that the stress, strain, conductivity and thermal stability of SF have been visibly enhanced, with the mechanical properties being comparable with those of literature reported super SF and even the spider fiber. The CNT is sheathed by silk protein and aligned within SF, and this would change the cross-section of SF from the triangle to ellipse. The CNT-embedded in SF caused the silk-I structure change due to the decrease in amide-II structure and the increase in amide-III structure.

7.2 Case on directly formation of magnetic silk fiber from silkworm

Magnetic responsive materials are the topic of intense research due to their potential breakthrough applications in the biomedical, coatings, microfluidics and microelectronics fields. By merging magnetic and polymer materials one can obtain composites with exceptional magnetic responsive features. Magnetic actuation provides unique capabilities as it can be spatially and temporally controlled, and can additionally be operated externally to the system, providing a non-invasive approach to remote control.

Magnetic materials can be applied to the security paper, health-care products, magnetic filters, and electromagnetic shielding.

Among various magnetic materials, the magnetic biomaterials are preferred because they have some novel application in medicine areas. As has been known, the SF with magnetic properties has been developed and reported by some researchers by taking the magnetic nanoparticle, Fe_3O_4, to coat the SF surface to form MSF. Additionally, the magnetic cellulose fibers, MSF, have been produced by using magnetite to coat, lumen loading, in situ synthesis within cellulose matrix, dispersion in cellulose solutions, and embedded using ionic liquids.

In this work we introduced a simple method to obtain pristine MSF by applying magnetic nanoparticles to feed silkworm together with the mulberry leaves then directly obtained MSF. The obtained MSF was compared with the normal SF both parallel fed in our lab. Additionally, the values of MSF obtained in this case were also compared with literature reported values.

In this work, ten B. mori larval silkworms were reared on an artificial diet at our laboratory. Two kinds of mulberry leaves, MLs, were prepared to feed to silkworms, one is the inartificial MLs, and another is the MLs spread with Fe_3O_4 powder to form the MMLs. After one month feeding, 10 silkworm cocoons were obtained. In this case, the used Fe_3O_4 powder is a commercial sample obtained from Guangzhou Jiechuang Trading Co., Ltd. China. According to this company, the size of these Fe_3O_4 powders is about 100 nm and a purity of about 99.8%.

The MSF and normal SF were collected by immersing the silkworm cocoons in warm water (about 80～90 ℃) with a pH about 9～10 for about 1～3 h then hand taken to wind in a glass bottle surface. Each SF was collected and prepared at about 10 m in length for analysis. Before measurement, the obtained SFs were oven dried at 90 ℃ for 24 h and the moisture was measured in a range of 3%～5%.

Figure 7-6 shows the macroscopic images of the silkworm fed with normal MLs or MMLs and their resulted cocoons, respectively. One can observe the magnetite-embedded cocoon that presented light black color as compared with the normal cocoon due to the embedded original Fe_3O_4 powder in dark color.

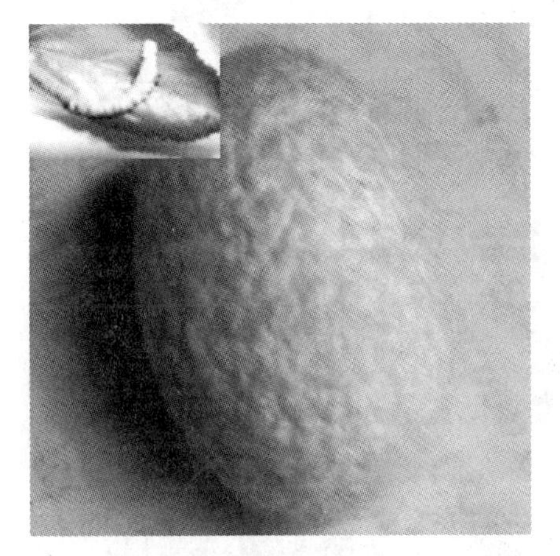

Feeding silkworm with only the mulberry leaves and yielded cocoon

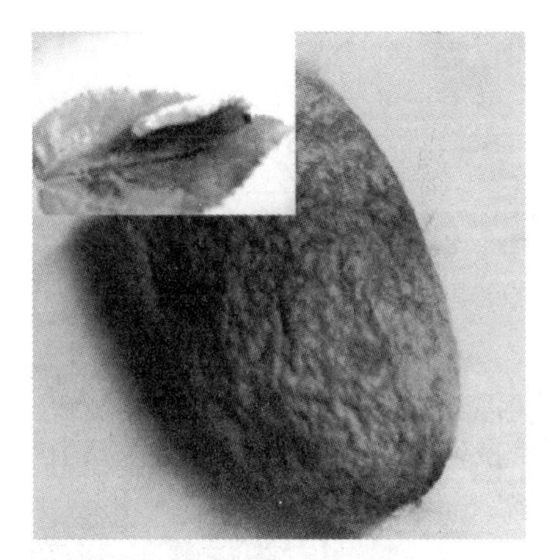

Feeding silkworm with nano Fe_3O_4 powders together with mulberry leaves and yielded magnetite-embedded cocoon

Figure 7-6 A comparison of the macroscopical images of the normal feeding of silkworm with only mulberry leaves and the yielded cocoon (left) and the feeding of nano Fe_3O_4 powders together with the mulberry leaves and the yielded magnetite-embedded cocoon (right)

The XRD patterns and SEM micrographs of these two SFs were compared in Figure 7-7. The presence of Fe_3O_4 in SF has been found to influence the crystalline structure of SF as shown in the XRD patterns in Figure 7-7 (top), because the comparison found that the SF exhibits the typical 2θ peaks at 9.5° (010), 20.8° (210) and 24.5° (002) (Figure 7-7, top left) ascribing to the β-sheet crystalline structure of silk fibroin, and these peaks were shifted at 9.2°, 20.9° and 23.5°, respectively, for MSF (Figure 7-7, top right). In the XRD pattern of MSF, a characteristic peak of Fe_3O_4 was found located at about 30.3° (220), to prove the magnetic materials embedded in SF. Since the SF-based peaks corresponding to the β-crystalline spacing distances and these peaks shifted upward or downward both in relation to the crystallite size reduce or increase, this not only proved that the Fe_3O_4 was embedded in SF but also indicated that this embedding changed the crystal structure of SF. SEM micrographs comparison showed that the SF has a smooth surface with visible grooves aligned along the fiber axis (Figure 7-7, left) and the MSF surface presented sort of debris and the grooves seem to be covered by the gum-like sericin proteins (Figure 7 - 7, right). These different morphologies suggested that the Fe_3O_4 is embedded in SF, and the presence of those inorganic materials within SF influenced the silkworm spinning process.

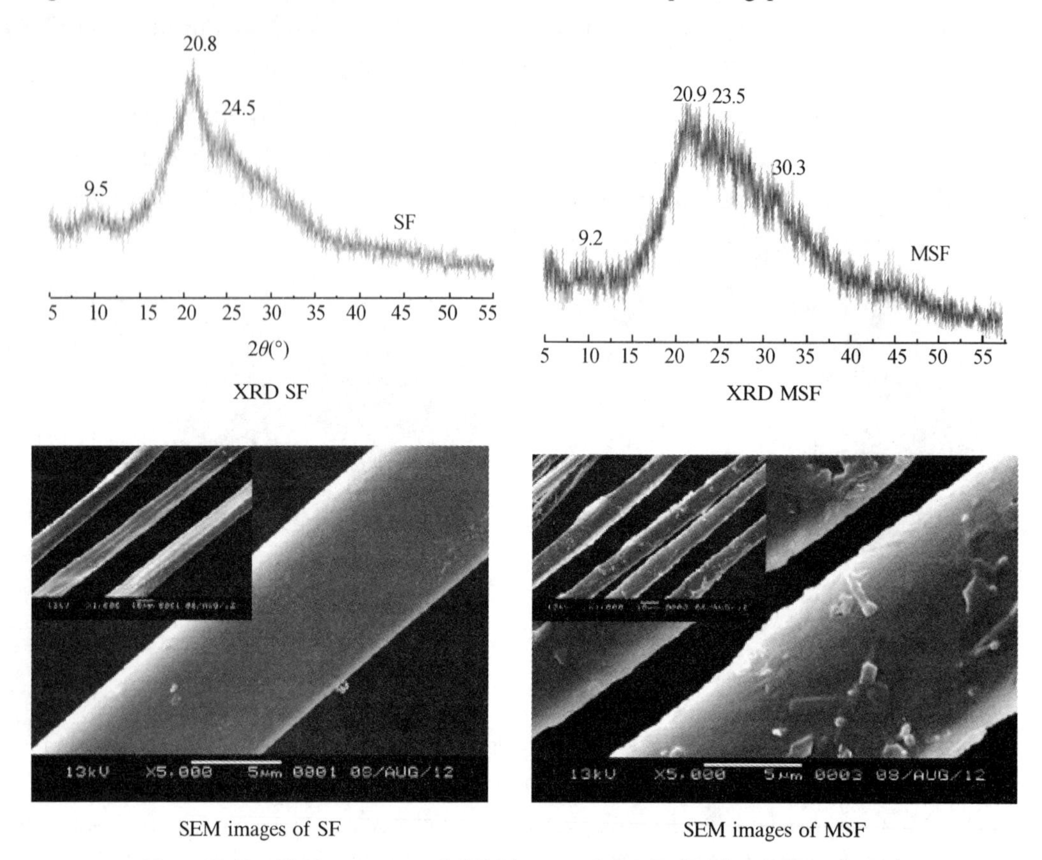

Figure 7-7 XRD patterns and SEM images of the SF (left) and MSF (right)

Table 7 - 3 quantitatively compared the differences of degree of crystallinity and orientation between SF and MSF. According to Table 7 - 3, the magnetite powder embedded in SF would reduce the crystallinity of SF about 32% and increase small degrees of orientation.

Table 7-3 Degree of crystallinity, DC, and degree of orientation, DO, of the SF and MSF

Silk fibers	Crystallinity (%)	Orientation (%)
SF	24.7	93.6
MSF	18.8	93.8

The fluorescent properties of these two SFs were compared. A broad excitation band extending from 250 nm to 450 nm was observed when the monitoring wavelength at 362 nm. A comparison of two samples found that the normal SF presented two intense peaks at 341 nm and 358 nm (Figure 7-8, top) both corresponding to the $\pi \rightarrow \pi^*$ electron transition, and these two peaks were shifted upward at 362 nm and 372 nm in MSF, respectively (Figure 7 - 8, bottom). This difference indicated that the magnetite embedded in SF without changed the main SF structure while the fluorescence properties were changed. The emission spectra comparison showed that the normal SF has an intense peak located at about 543 nm and this peak was hypsochromically shifted to 535 nm in MSF corresponding to the $n \rightarrow \pi^*$ transitions. The reason behind this shifting is due to the interaction between Fe_3O_4 and silk protein. This evidence indicated the presence of Fe_3O_4 powder in SF.

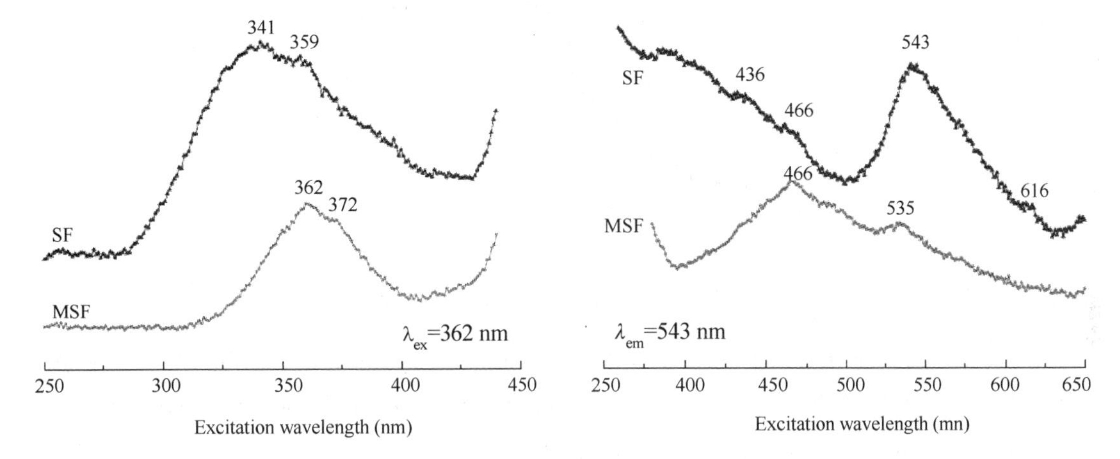

Figure 7-8 Fluorescence spectra of the SF and MSF

The magnetic property of MSF was studied and compared with the SF as shown in Figure 7-9. As expected, the normal SF without showed any magnetic response while the MSF presented a typical hysteresis loop to proven this MSF having effective magnetic property available responded by extra magnetic field. This indicated that the magnetic powders were indeed embedded in SF.

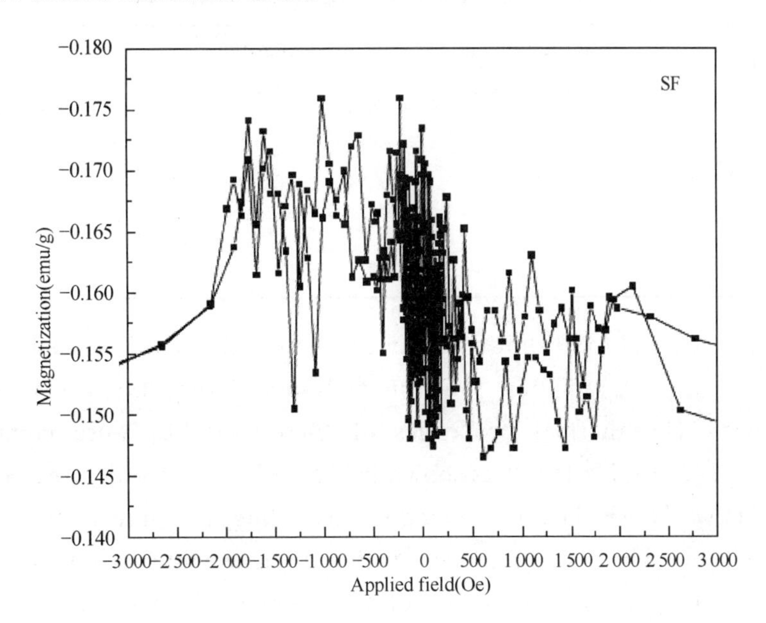

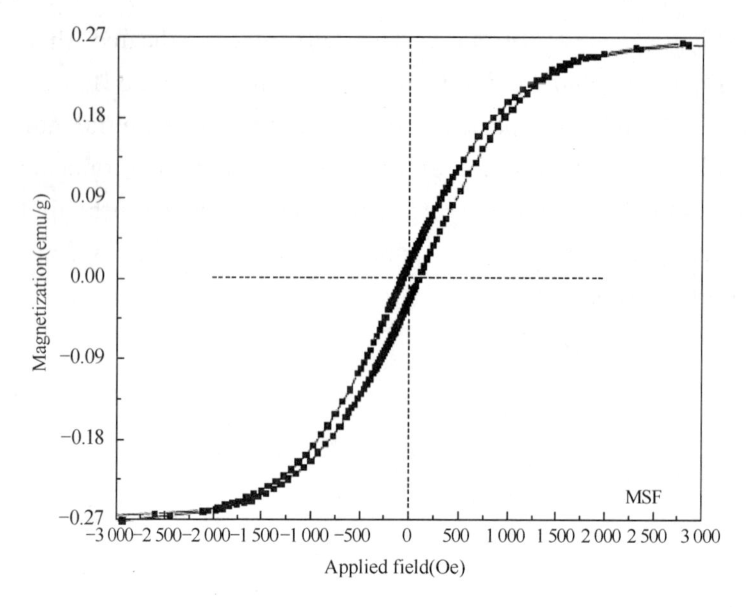

Figure 7-9 A comparison of the magnetic properties of the SF and MSF

Table 7-4 presented and compared several magnetic values on we obtained MSF and literature reported MSF and MCF. It was found we made MSF showed small magnetic values as comparison with literature reported relative values on MSF and MSF. These differences are caused by used different methods. For example, the MSF prepared by *Chang* et al. is used radiation to treat the silk protein solution containing the nanometer-scaled semiconductor crystallites, also known as the quantum dots (QDs), e. g. the $ZnAc_2$, $MnAc_2$ and SeO_2. While the MCF prepared by Sun et al. is using the ionic liquid to dissolve cellulose then mixed with Fe_3O_4 powder.

Table 7-4 Comparison of the magnetic properties of SFs

Samples	Saturated magnetic intensity (emu/g)	Retentivity (emu/g)	Coercivity (Oe)
SF	0	0	0
MSF	0. 3	0. 02	84
MSF	1. 5	0. 07	25
MCF	3. 7	0. 53	141

Though we prepared MSF showed lower magnetic values, it is availably applying to produce cloths with anti-radiation function and some biomedical cases such as drug delivery systems. The thermal properties of these two SFs were compared using thermogravimetric analysis (TGA) as shown in Figure 7-10, of which the derivative TG and DTG results were also shown in window. According to Figure 7-10, both samples have a little moisture which was removed in the initial weight loss stage, and the MSF has enhanced thermal stability than that of the normal SF. A comparison of the TG values of MSF and SF also suggested that the embedded Fe_3O_4 powder within MSF is about 5.9% in weight. This is found similar as MCF. In fact, the cellulose mixed with Fe_3O_4 powder has also been found enhanced thermal stability.

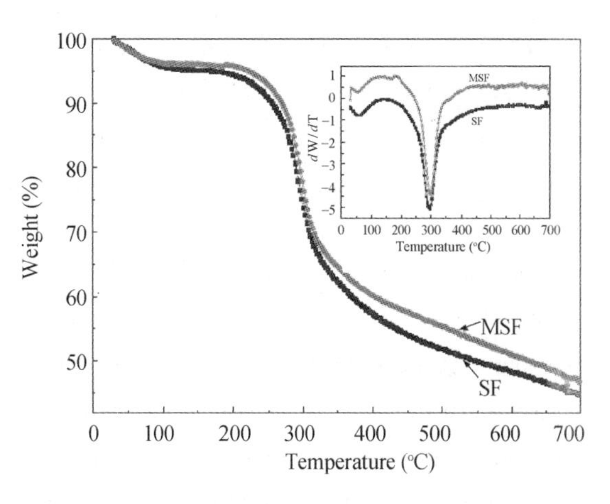

Figure 7-10 The TG curves of the SF and MSF

The mechanical properties of SF and MSF were compared in Figure 7-11, and it is clear that the MSF has better breaking stress than that of the normal SF. This suggested that the Fe_3O_4 powder would be embedded within the net structure of silk protein. In terms of Figure 7-11, it is, of interest, found that the fed Fe_3O_4 powder mixed with original silk protein caused increase of both the stress and strain of MSF because such mechanical behavior is less observed to indicate MSF has broad application ability than that of the normal SF.

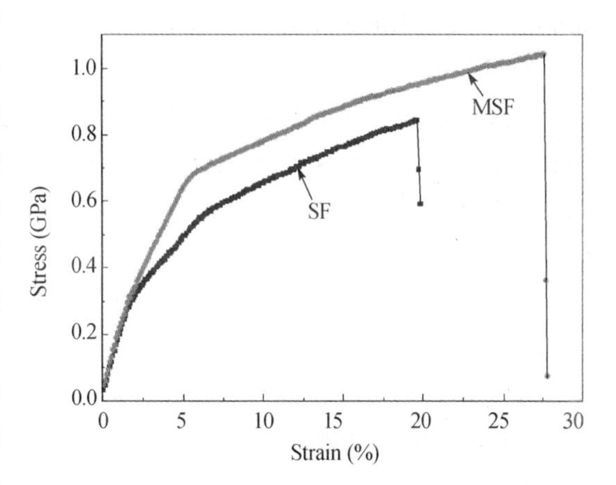

Figure 7-11 The mechanical properties of the SF and MSF

The experiment has proven that the pristine MSF can be directly obtained from silkworm by feeding the nanomagnetic powder to those worms. Compared with the normal SF, the obtained MSF has been found not only with expected magnetic properties, but also with enhanced thermal stability and mechanical properties. These advances indicated that one can use a simple method to obtain functional SFs to fit the request of some novel biomedical applications.

7.3 Case on formation of lotus root fiber

Natural dietary fiber has possibility in improvement of the health in the general populace due to its some physiological benefits. Such fibers are also known with excellent internal cleanser to absorb and remove harmful waste products and poisonous materials, reducing cholesterol and heavy metal levels, and to prevent and treat such ailments as diabetes, obesity, and cancer. Thus, such fibers have been strongly recommended even authorized by some countries. On dietary fiber, it is generally known that it vary in composition and structure and these depend on the plant origin, age, and converting method. If fact, it is also known that these variations of fiber are caused by different physiological effects of humans.

As one of hygrophilous plants and a member of Nymphaeaceae family, the lotus root has been long time known growth in warmly area especially native to Asia. This underwater-obtained aquatic plant is early taken as a food for Chinese and the time to be started since the Han Dynasty, e. g. about 207 B. C. to 220 A. D. Not considered very glamorous a fine vegetable with well-deserved culinary respect, the pierced lotus root has been also found to have air tunnels structure that looks something like large snowflakes or symmetrical rounds of a cheese. The nutrition facts of lotus root have been also known similarly as that of potato with low caloric components, e. g. abundant vitamins, minerals and polysaccharides. Additionally, literature also introduced that the lotus root is capably taken as a traditional medicine for therapying several diseases, especially for anti-HIV. However, in addition to broad reported application cases as mentioned above, the information on lotus root fiber is exactly less reported.

Taking a fresh lotus root as raw material purchased from a local market at Shanghai by washing several times with distilled water, the lotus root was hand-broken to reveal natural air tunnels structure where the lotus root fibers presented random. Since the natural lotus root fibers are presented desultorily at the broken surface of cross-section also including in the wall of tunnels, we use hand to draw and winded it on a glass bottle. The obtained lotus root filament was dried using an oven under a temperature of about 50 ℃ for several hours.

Figure 7-12 presented the SEM images of lotus root fibers. Observed that the lotus root fiber was juxtaposed by several filaments and the arrayed radixes from four to eight (Figure 7-12, A). Additionally, the lotus root fiber has been found to show regular nodes in initially (Figure 7-12, B), while these nodes were disappeared in Figure 7-12 (C) and (D) considerable due to the longer warming time for enlarging the SEM photographs. This suggests that these nodes are contributed by some fatty acids and the disappearance is caused by melting. According to Figure 7-12 (D), the averaged diameter of unique lotus root fiber is about 500 nm.

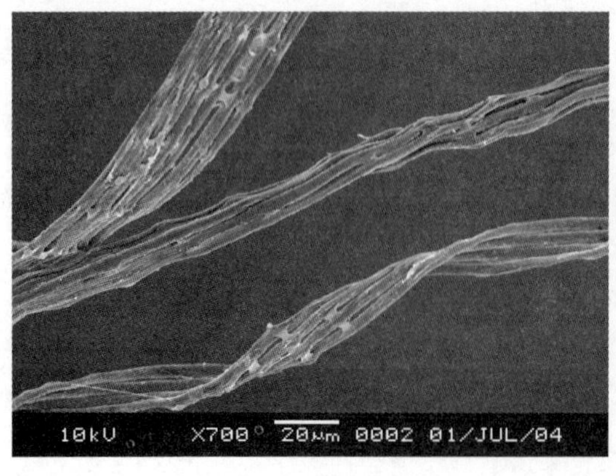

(A)

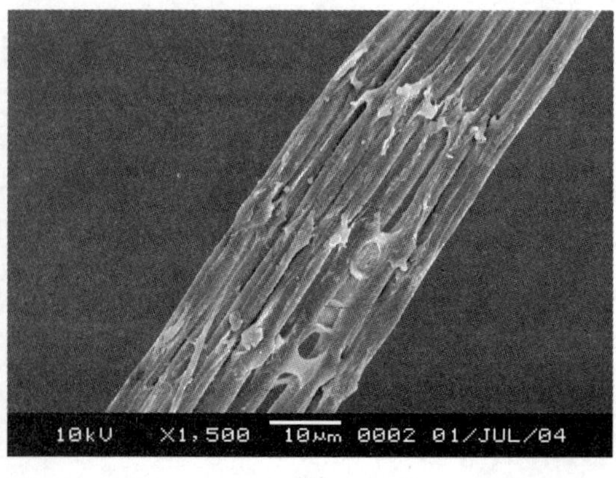

(B)

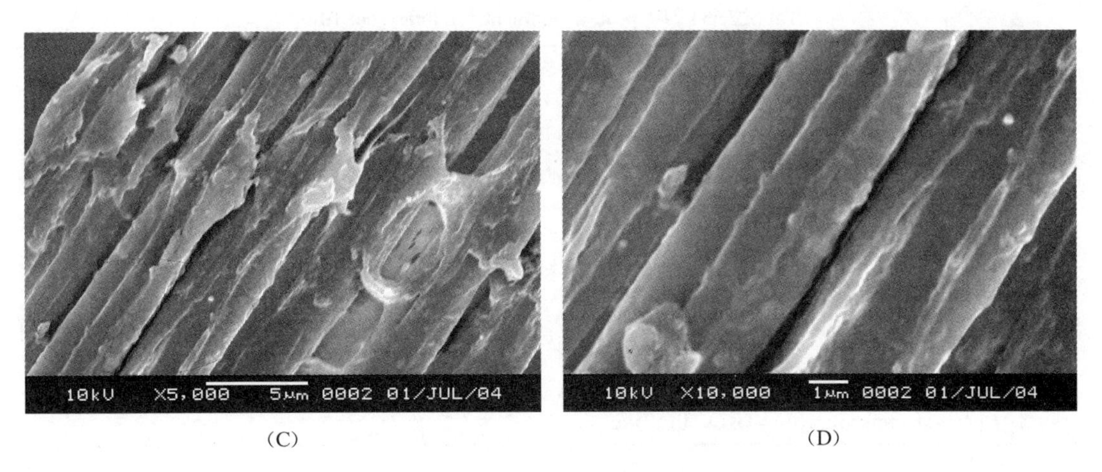

(C) (D)

Figure 7-12 SEM images of lotus root fiber

The breaking intensity and elongation of single lotus root fiber were measured at about 1.3 cN and 1.1%, respectively.

The thermal property of lotus root fiber was measured by DSC. Figure 7-13 suggested that lotus root fiber has a glass transform temperature, T_g, at about 43 ℃.

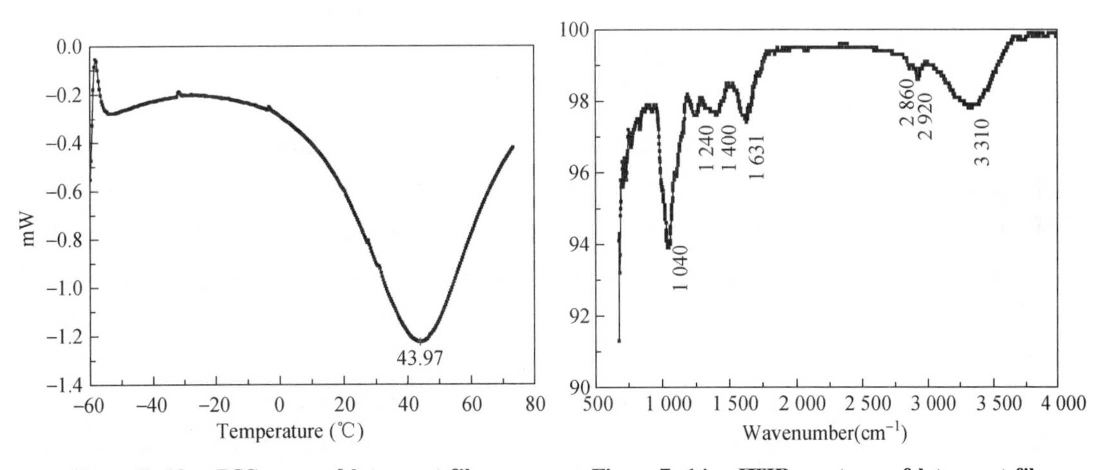

Figure 7-13 DSC curve of lotus root fiber **Figure 7-14 FTIR spectrum of lotus root fiber**

FT-IR spectrum of lotus root fiber was presented in Figure 7-14. The visible intense peaks were assigned and summarized in Table 7-5. According to Table 7-5, it is generally known that lotus root fiber is composed by protein and lignocellulosic components.

This work proven that the natural lotus root fiber was availably obtained. Results showed this dietary fiber composed of polysaccharides, lignin, fatty acid and protein as most plants. The averaged diameter of a single fiber is about 500 nm with certain mechanical properties. This fiber has a T_g at about 43 ℃.

Table 7-5 IR peak assignment for lotus root fiber

Wavenumber (cm^{-1})	Components	Wavenumber (cm^{-1})	Components
1 040	cellulose, polysaccharides	2 860	OCH_3 absorb peak, fatty aldehyde
1 240	C—O libration, gelose, hemicelluloses	2 920	CH_2 stretch vibration, polysaccharides, lignin
1 400	—CH_2—CO—, fattiness	3 310	R—NH—R stretching vibration, amino group
1 631	extended peptide chain, NO_2 stretch vibration, H_2O vibration		

7.4 Case on formation of herb fiber

7.4.1 Case on formation of persimmon leaves-based fiber

Since persimmon leaves, PL, have been found to have beneficial effects on haemostasis, constipation and hypertension, they have been broadly applied in food and medicine area, especially recently. Therefore, it is a consideration of us to convert this material to a fiber to extend its application area.

7.4.1.1 Formation of cellulose/persimmon leaves blending fiber

Recently, studies of PL have been carried out in our laboratory, and their contents and several properties, e.g. surface and dynamic dissolution in two solvents, e.g. DMSO and DMAc, have been characterized. These studies provided results that allowed us to prepare a PL/cellulose blend fiber by means of wet spinning as aimed. A reference cellulose fiber was also prepared using the same method and condition as that of blend fiber.

The used PL was picked from Shanghai and prepared as powder with a size of about 40 meshes. The cellulose used was a commercial product provided by Shanghai Chemical Fiber Co. and known to have degree of polymerization (DP) of about 500, and α-cellulose content about 90%.

Polyoxymethylene (PF) and DMSO were analytical grade solvents purchased from a local chemical company (Shanghai) were used as received without further purification.

The formation of cellulose/PL blending fiber was performed by weighing the amounts of PL (2%wt~10%wt), cellulose (10%wt) and DMSO/PF(10%) were used to prepare spinning solutions. Solutions were initially stirred and heated to 60 ℃ held for 1 h, then heated to 110 ℃ and held until that temperature transparent. The residual formaldehyde within solution was removed using a vacuum pump at 80 ℃ for several hours. Before spinning, this solution was stirred again and filtered to remove insoluble solid materials.

Laboratory scale wet spinning was employed. The process was similarly as that of *Focher* et al. Under a pressure of about 30 kPa and controlled by a metering pump, e. g. 0. 2 g/min, spinning solution was allowed to flow into a 100 μm spinnerette die, *L*/*D* ratio of 1. 0, to form a persimmon leaves/cellulose blend filament. Then, this filament was passing and solidified by through two water baths. The first was a coagulation bath with a temperature of about 40～60 ℃; take up rate, V_1, was about 1. 2 m/min. The second was a boiling water bath; take up rate, V_2, about 2. 4 m/min. Finally, the filament was collected on a winder at a rate of about 10 m/min. After that, the filament was washed with distilled water, and then dried in a vacuum oven at 50 ℃ for 24 h. Before analysis and characterization, the filament was held at 20 ℃ and 65% relative humidity.

A cellulose filament was also prepared using the same conditions as a reference.

To apply wet spinning to prepare a filament, the influence from the temperature of coagulation bath was initially investigated. As Table 7-6 reports, a referenced cellulose filament was prepared. Of this Table 7-6, we found that both the lowest and highest temperatures are unacceptable for preparation of fiber. The data in Table 7-6 also suggests that the best temperature for coagulation bath seems to be in the range of 40～50 ℃.

Table 7-6 Influence of the temperature of coagulation bath on preparation of referenced cellulose fiber

Temperature of coagulation bath (℃)	Fiber drawing behavior	Fiber breaking intensity (cN/dtex)
20	Unable to draw	—
30	Breaks easily	1. 6
40	Good drawing	2. 1
50	Good drawing	2. 2
60	Breaks easily	1. 9
65	Unable to draw	—

To produce a PL/cellulose blend fiber, understanding of this material's compositions is generally required because they are usually in variety probably to influence blend fiber. Since the percent of PL added to the blend fiber was expected high for bioactivity, so the bio-fiber prepared with three different percents of PL.

Table 7-7 indicates that the persimmon leaves affect the orientation and mechanical strength. This is reasonable because rigid polymer cellulose is mixed with a random coil, low molecular weight matrix. According to Flory, poor compatibility occurs upon blend polymers beyond a critical concentration in the absence of strong intermolecular interactions based on the interference of the random coil polymer with a mutual orientation of the molecules of the rod-like polymer molecules. In fact, in this case the low-molecular-weight polymer of PL seemed to have acted as a plasticizer facilitating rotating of polymer chains. The same phenomenon was also observed in a case of wood fiber preparation by *Focher* et al. Based on Table 7-7, the decrease in mechanical properties seems to be less than expected from the amount of PL added.

Table 7-7 Influence of the amount of persimmon leaves, PL, added on the mechanical properties of PL/cellulose blend fibers

Fibers	Added PL amount (wt%)	Fiber orientation (%)	Fiber breaking intensity (cN/dtex)
Cellulose	0	70.0	2.20
PL/Cellulose blend	2	69.0	2.10
PL/Cellulose blend	5	67.5	2.00
PL/Cellulose blend	10	64.5	1.95

By fixing the percent of PL added, e.g. 5%wt, and varying the concentration of spinning solution from 6%wt～12%wt, its influence on fiber preparation and mechanical properties of the blend fiber was also investigated. As summarized in Table 7-8, a lower concentration, e.g. 6%, seems to be inappropriate for preparation of fiber. Moreover, the mechanical properties of the blend fiber increased with an increase of the concentration of spinning solution, in good agreement with the above-mentioned Flory theory.

Table 7-8 Influence of the spinning solution concentration on preparation and mechanical properties of persimmon leaves/cellulose blend fiber, PL-persimmon leaves

Concentration of spinning solution (%wt)	Added PL amount (%wt)	Fiber drawing behavior	Fiber breaking intensity (cN/dtex)
6	5	Difficult drawing	—
8	5	Breaks easily	1.8
10	5	Good drawing	2.0
12	5	Good drawing	2.2

Table 7-9 compares X-ray data of the reference and blend fibers resulted from Figure 7-15. From Figure 7-15 and Table 7-9, it was observed that the blend fiber crystalline was less difference, e.g. mainly displayed at $10\bar{1}$ and 102 peaks comparing to reference cellulose fiber.

Table 7-9 A comparison of the crystallinity and crystal grain size for persimmon leaves/cellulose blend fiber and referenced cellulose fiber

Fibers	Added PL amount (%)	Crystallinity (%)	Crystal grain size (nm)		
			101	$10\bar{1}$	102
Cellulose	0	69.6	1.17	1.66	1.19
PL/Cellulose blend	10	66.5	1.17	1.61	1.23

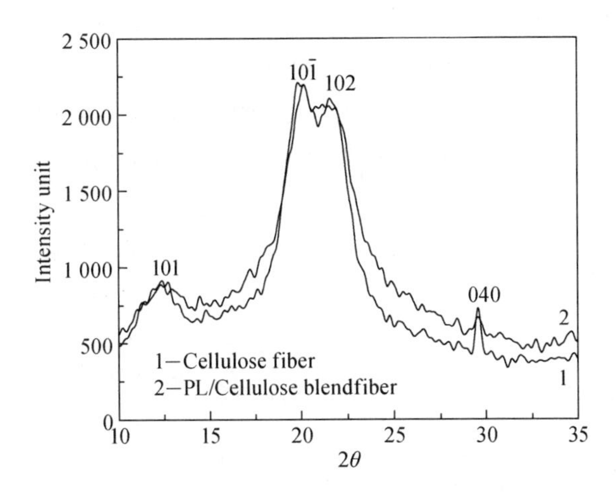

Figure 7-15 A comparison of X-ray diffractograms for referenced cellulose fiber (1) and persimmon leaves/cellulose blend fiber (2)

Figure 7-16 showed DSC curves for these two fibers. A comparison of these curves found that the glass transition temperature, T_g, of the blend fiber is about 4 ℃ less than that of reference cellulose fiber. For comparison, Figure 7-17 is the curve of persimmon leaves alone.

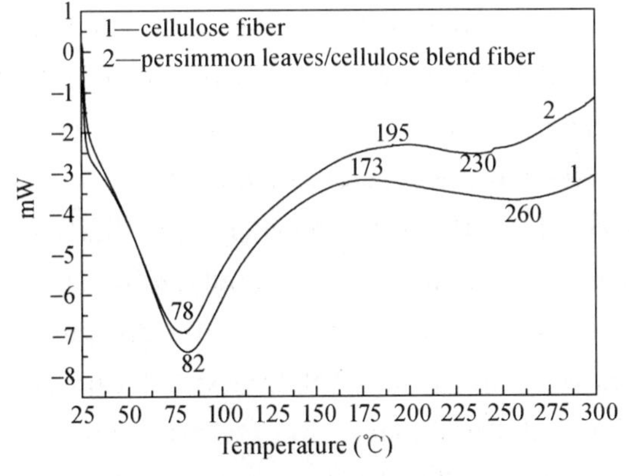

Figure 7-16 A comparison of DSC curves for referenced cellulose fiber (1) and persimmon leaves/cellulose blend fibers (2)

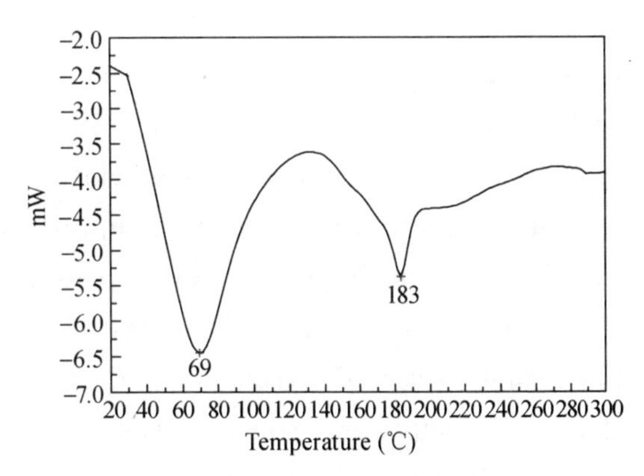

Figure 7-17 DSC curve of persimmon leaves

This work proven that the use of DMSO/PF solvent and wet spinning one can prepare herb fibers.

Since this PL-based fiber is effectively in haemostatic, dieresis, constipation and hypertension, to study the drug release behavior is wished and also investigated.

Taking measured conductivity as a function for release time, Figure 7-18 described a dynamic drug release process for we prepared PL/cellulose blending fiber. A comparison with referenced cellulose fiber, e. g. both corresponding to the same temperature at 25 ℃ (Figure 7 - 18), showed that the PL/cellulose blending fiber presented good drug release possibility due to its conductivity values greater than that of reference indicating that a fiber blending with drug component is capably for recommending to medical area as a new drug delivery system.

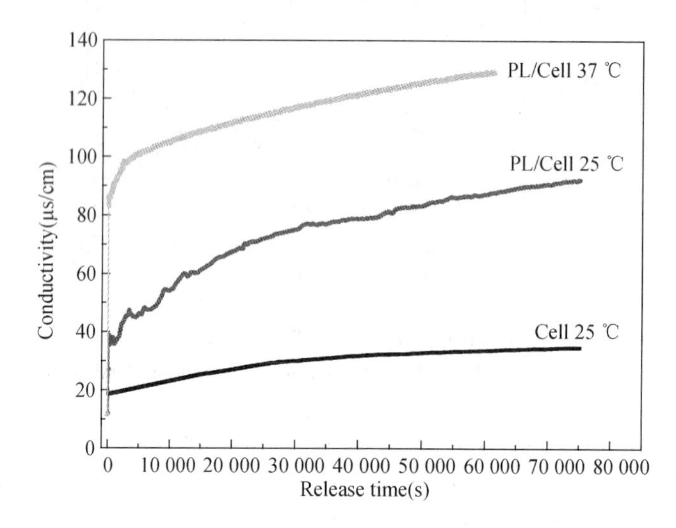

Figure 7-18 Influence of temperature on dynamic release behaviors for PL/cellulose blending fiber and referenced cellulose fiber

According to Figure 7 - 18, it is also observed that drug release was generally controlled by temperature. This is because that the release curves clearly showed: the higher the temperature, the greater the release occurrence.

7. 4. 1. 2 Formation of poly(acrylonitrile)/persimmon leaves blending fiber

In the process of spinning solution preparation, the PAN powders were initially heated to increase its temperature up to about 80 ℃, and kept for about 12 h. Then, PAN powders were dissolved in DMAc solvent for about 3～4 h until fully dissolution phenomena observed. After that, PL powders were added to PAN/DMAc solution and the triple components solution was further heated for about 1 h to yield a uniform solution. Before spinning, the blend solution was filtrated to fit the request of spinning.

Briefly, the fiber was spun under a pressure of about 300 kPa controlled by a metering pump with a flow rate of about 0. 2 g/min. The fiber was initially formed by a spinneret with the diameter of 100 μm and a L/D ratio of 1. 0, then be solidified by thronging out two water baths, i. e. the former was a coagulation bath with a temperature of about 55 ℃, and the latter was a boiling bath with a temperature of about 95 ℃. Finally, obtained filament was washed by distilled water and dried by a vacuum oven at 50 ℃ for 24 h. Before analysis, the filament was held at a container under a condition of 20 ℃ and 65% relative humidity.

As same, a pure PAN fiber was prepared as a reference.

In general, to blend polymeric material with PL is based on a known fact that this natural material is rich in macromolecules components. For example, by analysis of its material components, it was primarily known that PL consists of cellulose about 68%, hemicellulose about 8% and lignin about 12%, respectively. However, in the case of PL/cellulose blend fiber, it was exactly found that the effects from the low-molecular-weight components and anomalous macromolecules of PL are obviously on the structure and mechanical properties of obtained blend fiber. Additionally, influences from spinning process, e. g. the temperature of coagulation bath and the concentration of spinning solution, are visibly. Furthermore, the amount of PL added is also an important factor.

To vary the temperature of coagulation bath from 20 ℃ to 50 ℃ and take fiber drawing and breaking intensity as two targets, the influence of coagulation bath on preparation of PL/PAN fiber was evaluated in Table 7-10. Observe, both the lowest and the highest temperatures, e. g. 20 ℃ and 50 ℃, respectively, caused bad results indicating these temperatures are un-capable for applying to preparation of the PL/PAN blend fiber. According to Table 7-10, the best temperature of coagulation bath seems to be the middle, e. g. about 40 ℃. Since this temperature is smaller than that for preparation of PL/cellulose blend fiber previously adopted, this suggests that the solidification of PL/PAN blend fiber is easily than that of PL/cellulose blend fiber. Probably, this may benefit of the enhancement of the mechanical properties for PL/PAN blend fiber.

Table 7-10 Influence of temperature of coagulation bath on drawing and mechanical properties of PL/PAN blend fiber

Temperature of coagulation bath (℃)	Fiber drawing behavior	Fiber breaking intensity (cN/dtex)
20	Breaks easily	1.6
30	Good drawing	2.1
40	Good drawing	2.2
50	Breaks easily	1.9

By fixing the amount of PL added at about 5%wt and varying of the concentration of spinning solution from 14%wt~18%wt, the influence of the concentration of spinning solution on fiber was evaluated in Table 7 - 11. Observe that the lower concentration of spinning solution, e. g. 14% and 15%, are inappropriately for PL/PAN blend fiber preparation due mainly to difficult drawing.

According to Table 7-11, it was noted that the mechanical properties of PL/PAN blend fiber seems to be increased with the increase of the concentration of spinning solution especially up to about 18%. This is important and generally expected for preparation of a qualified polymeric filament.

Table 7-11 Influence of the concentration of spinning solution on drawing and mechanical properties of PL/PAN blend fiber

Concentration of spinning solution (wt %)	Added PL amount (wt %)	Fiber drawing behavior	Fibers breaking intensity (cN/dtex)
14	5	Difficult drawing	—
15	5	Breaks easily	1.8
16	5	Good drawing	2.1
18	5	Good drawing	2.2

To prepare a PL/PAN blend fiber, the amount of PL added is expected in high. However, the amount added is considerable to influence polymer's structure and to influence related mechanical properties. The reason is due to PL with various low-molecular-weight substances that would influence the polymer net and extending to the orientation of fiber. In order to understand the amount of PL added is how to influence blend fiber, the PL added in spinning solution was thus varied from 0, 2.0%, 5.0% to about 10% in weight, respectively. Of which, the zero represents the referenced PAN fiber.

According to X-ray measurement presented in Figure 7-19, the influence of the addition of persimmon leaves on PAN is exactly as the evidence that the intense peaks visible shifted. Moreover, the visible change seems to be at the 100 and 101 peaks for blend fiber due to the former smaller and latter greater in comparison of PAN fiber (Figure 7-19). This thus indicated that the crystal structure of PAN was changed by adding of PL. Figure 7-20 further showed that the crystallinity of PAN was rapidly decreased with the increase of the amount of PL added. As a support, Figure 7-21 showed that the orientation of blend fiber was greatly reduced with the increase of the amount of PL due to the change of crystal structure. Figure 7-22 furthermore described

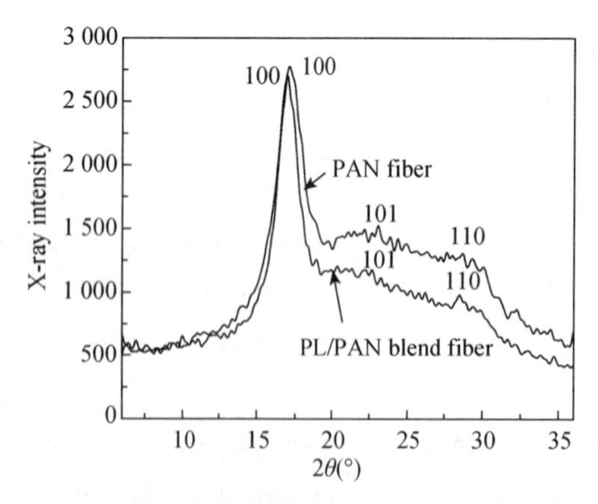

Figure 7-19 A comparison of the X-ray diffractograms for PL/PAN blend fiber and referenced PAN fiber

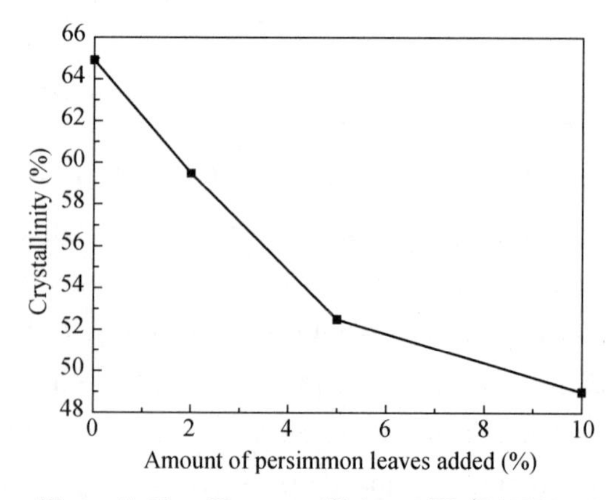

Figure 7-20 The crystallinity of PL/PAN blend fiber as a function of the added amount of persimmon leaves

that the change of the crystal structure is indeed to influence the mechanical properties of blend fiber. This suggests that the preparation of PL-based bio-fiber needs to choose a suitable concentration to fit the request of remain either the bio-component or the mechanical properties.

Based on Figure 7-20, the change of the crystal structure is detailed described in Table 7-10. Observe that the change of the crystal structure is occurred at the crystal grain size. This thus furthermore indicated that the low-molecular-weight substances remained in PL would cause poor compatibility for blend polymer where beyond a critical concentration in the absence of strong intermolecular interactions based on the interference of the random coil polymer with a mutual orientation of the molecules of the rod-like polymer molecules. This is possible and has been ascribed to that the low-molecular-weight polymers played as a plasticizer to facilitate polymer chains.

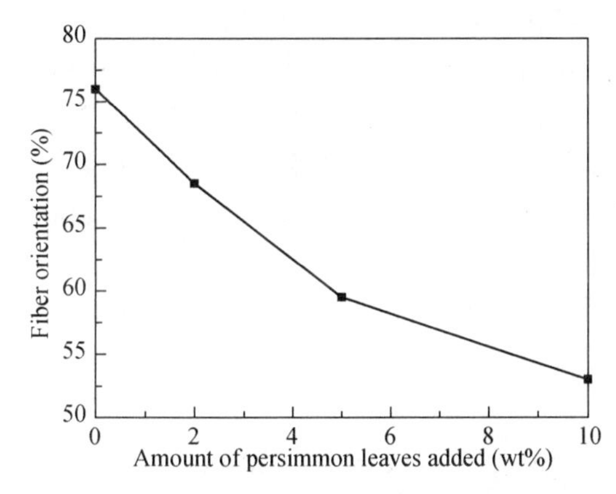

Figure 7-21 The orientation of PL/PAN blend fiber as a function of the added amount of persimmon leaves

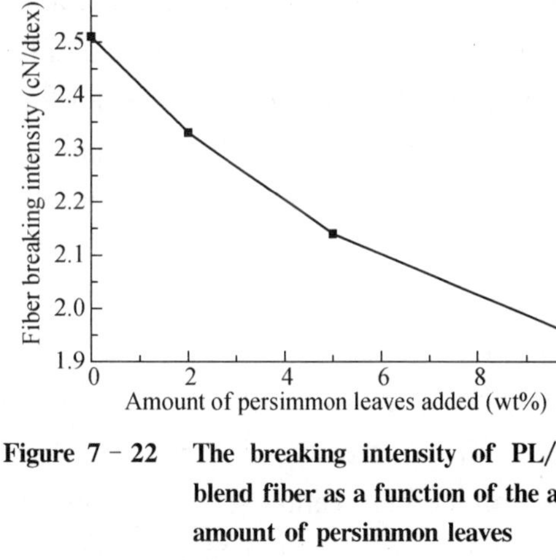

Figure 7-22 The breaking intensity of PL/PAN blend fiber as a function of the added amount of persimmon leaves

In order to understand the difference between PL/PAN blend fiber and PAN fiber, DSC curves for these two fibers were recorded and presented in Figure 7-23. A comparison of these curves showed that the glass transition temperature, T_g, for blend fiber is visible reduced at about 72 ℃ than PAN fiber at about 95 ℃. Furthermore, the comparison of the crystallization temperature, Tc, for both blend and pure fibers however showed that it seems to be similarly without influenced. This is of interest since Figure

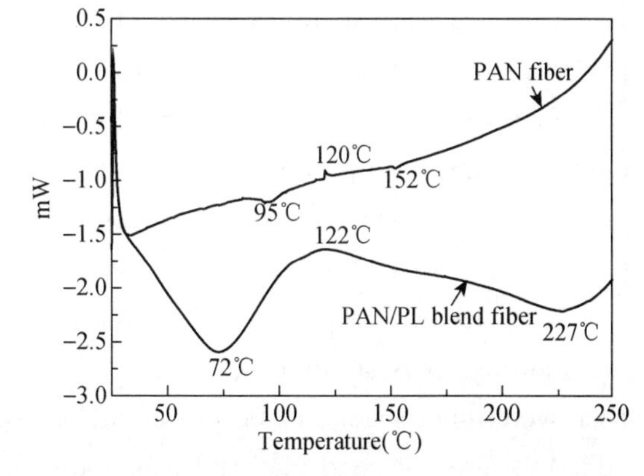

Figure 7-23 A comparison of DSC curves for PL/PAN blend fiber and referenced PAN fiber

7-23 presented thermal phenomena seems to be on the contrary of Figure 7-5 showed information. Thus, it is considered that the addition of PL into PAN is acceptable for fabrication of a bio-blend fiber.

This work proven that the applying wet spinning method, a PL/PAN bio-blend fiber can be fabricated. According to structure analysis and mechanical properties determination, as well as compared to a referenced PAN fiber prepared via the same method in this case, it was found that this bio-blend fiber is capable for further application though the change in the crystal structure seeming to be visible due to the addition of persimmon leaves induced also several low-molecular-weight substances.

7.4.2 Case on formation of ginger-based fiber

Ginger (*Zingiber officinale Roscoe*) is a typical food-based drug recognized by Chinese and Chinese culture-based area. Hereby we reported a case for formation of ginger-based fiber by considering it belongs to the Zingiberaceae family and the rhizome of ginger species (*Zingiberaceae*) including ginger (*Zingiber officinale Roscoe*) and turmeric (*Curcuma longa L.*) has been long time regarded as an ethno medicine widely used as a traditional oriental medicine to ameliorate such symptoms as inflammation, rheumatic disorders and gastrointestinal discomforts.

In this case, dried ginger purchased from local traditional medical store at Shanghai was employed as drug. Before use, the dried ginger was crushed as powder with particle size of about 40 mashes.

Polyethylene glycol (PEG) with a molecular weight of about 4 000 purchased from local chemical company was directly applied without further purification.

A commercial cellulose with degree of polymerization, DP, of about 500 was employed as received from Shanghai General Pulp Factory without further treatment. According to analysis by producer, this cellulose is available for converting to filament due to its α-cellulose component above 90%.

Two solvents, e.g. polyoxymethylene, PF, and dimethy sulphoxide, DMSO, were employed and both purchased from a local chemical company at Shanghai with analytical grade. In this case, these two solvents were directly used as received without further purification. In addition, distilled water was always used through whole process.

In order to keep drug fiber remaining bioactivity as expected, a wet spinning methodology was adopted. A solvent blend, DMSO/PF(10%), was initially prepared, then weighted ginger, PEG and cellulose were blended with solvent. In this case, the added weight for each material was about 5.6 g for ginger, 10 g for PEG and 30 g for cellulose.

After blending with all solid materials, the solution was stirred and heated to 60 ℃, after remaining this temperature for about 1 h, then this blend was furthermore heated to 110 ℃ and to keep it until a transparent behavior observed. After that, residual formaldehyde probably remained within solution was extracted out by means of a vacuum pump at 80 ℃ for lasting several hours. Before spinning, stirring of this prepared solution was always kept. Finally, a filtration process was applied to remove insoluble solid materials, e.g. less than 0.1%wt.

Fiber wet spinning was carried out in a laboratory scale system under a pressure of about 30 kPa and controlled by a metering pump with a flow rate, e.g. about 0.2 g/min, the spinning solution was allowed to flow into a 100 μm spinnerette die with a ratio of L/D about 1.0 to form filament. Subsequently, this as-spun filament was quickly guided into two water bathes for solidification. Designed, the first was a coagulation bath with a temperature of about 40～60 ℃ and the second was a boiling water bath with a high temperature to be about 95 ℃. Finally, a winder with a preset rate of about 10 m/min was employed to collect filament. After that, this as spun filament was washed by distilled water then dried in vacuum oven at 50 ℃ for at least 24 h. Before analysis, this filament was kept at a glass desiccant at 20 ℃ and 65% relative humidity conditions.

In order to understand drug release behavior for prepared blending fiber, a pure cellulose fiber was also prepared as the same as blending fiber and taken as a reference.

In order to understand prepared ginger/PEG/cellulose blending fiber, FTIR spectroscopy was initially employed to record spectra of blending fiber as can be seen in Figure 7-24. The band assignment was summarized in Table 7-12. According to Figure 7-24 presented similar spectra and Table 7-12 reported peak height values, it was truly found that the structure of blending fiber without evidently changed after drug release. This suggests that this blend process only physical interactions taken place. In fact, since the IR spectra for both PEG and cellulose are available to find elsewhere, a comparison of their IR spectra with Figure 7-24 also support above mentioned suggestion. This

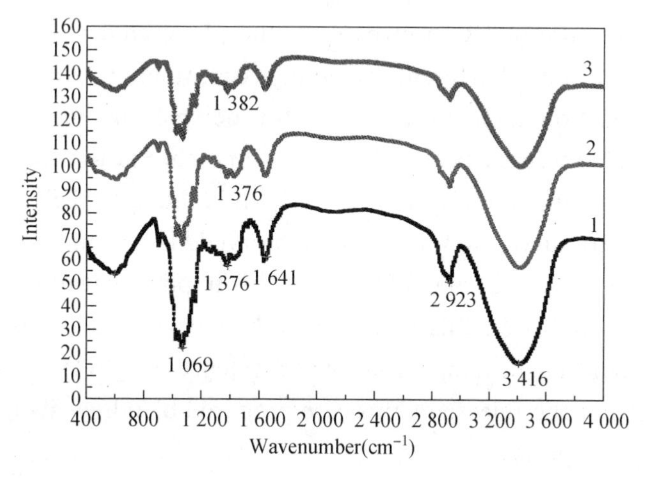

Figure 7-24 FTIR spectra of ginger/PEG/cellulose blending fiber recorded from different release processes, e.g. before release (1), during release (2) and after release (3)

conclusion has been found in good agreement with literature. From Table 7-12 reported values, it was also found that all presented intense peaks seem to be to reduce their heights when drug release occurrence. This means that IR spectra may also be taken as a tool for analysis of drug release quantitatively. Of interest, assuming that these values (Table 7-12) obtained from IR spectra could be taken as a measure, it is thus probably to estimate the amount of residual drug being remained within fiber. In other words, based on the fact that each peak is related to a structure of blending fiber in relation to employed drug, it is considerable to suggest that we may estimate structure-based drug components released from Figure 7-24. Certainly, this is sometime of importance for understanding drug release. Based on Figure 7-24, it seems to be a truth that the peak previously located at about 1 376 cm^{-1} for blending fiber without release has been moved to about 1 382 cm^{-1} corresponding to blending fiber after release. It is thus assumed that this peak reflected ginger combined with polymers and this structure closely in relation to release occurrence.

Table 7-12 Band assignments for IR spectra of blending fiber and a comparison of the release effect for blending fiber without release, during release and after release

Peak position (cm^{-1})	Band assignments	Peak height for sample 1 (%)	Peak height for sample 2 (%)	Peak height for sample 3 (%)
598	—	100	74	56
1 069	S=O stretch	100	88	63
1 376	Ring stretch	100	69	60
1 641	C=C stretch	100	90	55
2 923	CH_2 stretch	100	65	43
3 416	H_2O influence	100	85	62

In addition to the use of IR spectroscopy, the phase change behavior for blending fiber was measured using Different Scanning Calorimeter. In general, DSC can determine glass transition temperature, T_g, for polymer by increasing temperature. Moreover, based on presented peak related T_g, one can easily understand the rubber state or glassy state because the former above T_g while the latter below T_g. According to Figure 7-25 presented curve, it was clearly observed that a unique peak presented in reasonable ascribed to T_g, e.g. about 56.5 ℃. This implies that prepared blending fiber will change its phase based on this temperature boundary. Additionally, this indicates that a change of environment temperature will influence drug release behavior. In other words, this means that the release of this case prepared blending fiber will be controlled by its phase change property.

Following the DSC analysis, drug release experiments were performed and the results were presented in Figure 7-26. Of interest, it was evidently observed that temperature was an important factor to influence release occurrence, and this influence

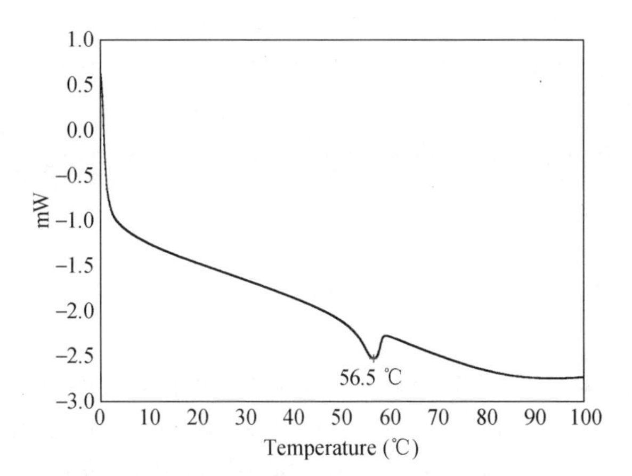

Figure 7-25 DSC curve of a ginger/PEG/cellulose blending fiber indicating its glass transition temperature at 56.5 ℃

seems to follow a regulation, e.g. the higher the temperature, the greater is the drug being released. Moreover, as expected that this Figure 7-26 clearly showed that a phase change was taking place because the temperature up to 65 ℃, absolutely greater than that of T_g, exactly causes the drug release rate decreased comparing to release below T_g. Without doubt, Figure 7-26 presented controlled drug release behavior verifies that a fiber-based release system is capably prepared.

A comparison of blending fiber with referenced cellulose fiber (Figure 7-26) further indicated that common fiber could not present release behavior than drug blend fiber though under the same temperature condition, e.g. 25 ℃. This again indicates that a fiber-based drug delivery system can be successfully prepared.

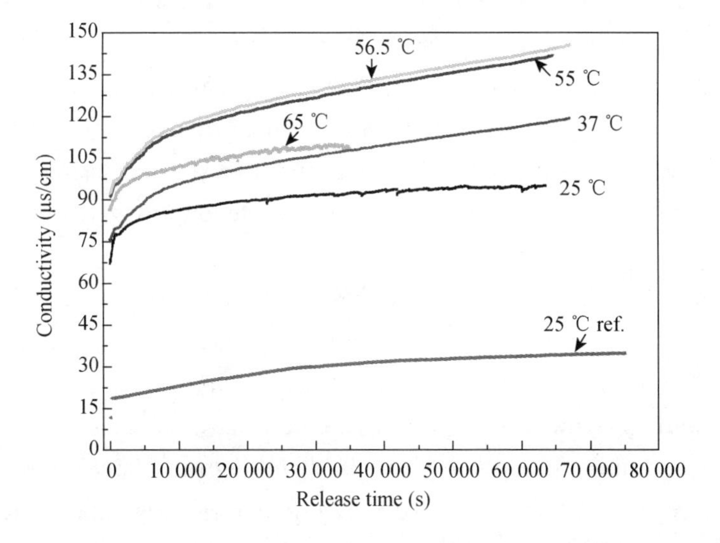

Figure 7-26 Drug release influenced by temperature and the phase change controlled release behavior for ginger/PEG/cellulose blending fiber

With respect to the truth that a fiber can be prepared with long length, it is thus assumed that we may apply fiber-based drug delivery system to some case where such shape is expected. Moreover, based on Figure 7-26 presented release behavior, it is also considered that the fiber-based drug release system may suit for applying to some cases where temperature sensitively.

7.5 Case on formation of chitin fiber

During the past decades, chitosan, CS, has got great attention and been broadly applied in medical and other areas with respect to its excellent biological characteristics of unique antiphlogistic effect, bio-compatibility, absorptivity, non-hypersensitivity, biodegradation property and wound healing. As known, CS has been currently used as scaffold for hepatocyte attachment, bacterial antigens, antimicrobial finishing reagents and capsules. Because CS has been also found available to accelerate the healing of human wounds, it was thus observed that this natural polymeric material has been thus converted into fiber as reported elsewhere. Obviously, this is an interest because fiber can play a role different to other forms. However, it is noted that CS fiber has low strength property, especially in wet statue. To overcome this shortage, some efforts have been therefore tried. For example, *East* and *Qin* have tried to acetylate CS fibers to enhance its mechanical properties, and *Knaul* et al. have applied two methods, i.e. one is the drying and another is the cross-linking, to improve the mechanical properties for CS fiber. According to these authors, an aqueous solution containing glutaraldehyde and glyoxal may be available to enhance the mechanical properties for CS. Noted, *Jegal* and *Lee*, *Moon* et al. and *Shin* and *Ueda* have tried to use aldehyde as reagent to cross-link CS fibers and membranes.

Since a high strength CS fiber is expected by medical area for applying to different cases such as bone settling, human body tissue, medical paper, wound dressing, anti-fungus fabric textile and others, we have prepared CS fiber using the cross-linking technique. In addition to *Knaul* et al., in this case we employ glyoxal as a cross-linking reagent.

The used powder CS was with a viscosity at about 625 mPa·s and degree of deacetylation at 91.2% and a glyoxal solution (30%) as cross-linking reagent. By dissolution of CS powder, 3.5% w/w, in aqueous acetic acid (4% V/V) to form a spinning dope, and applying a wet-spinning method.

To avoid cross-linking process resulting in shrinking for fiber, the prepared CS fiber was initially reeled in a "U" type glass rod then immersed in glyoxal solution to start the cross-linking reaction. Since cross-linking reaction is considerable to be influenced by many factors, an orthogonal experiment was designed as Table 7-13 described by taking the tenacity as a unique evaluation target and four vary parameters, e.g. time, temperature, pH and concentration of reagent. From 16 independent entries, an optimum condition was resulted in and applied for this case. For example, the better condition for cross-linking reaction has been found and applied as that of preparation of the cross-linking reagent with a concentration of about 2%, pH of 3.73, reaction temperature about 40 ℃ and reaction time 30 min. After cross-linking reaction, CS fiber was washed by distilled water for several times then dried in an oven condition, e.g. 50 ℃ for 16 h.

Table 7-13 A design of orthogonal experiment

Entry	Cross-linking reagent (%)	pH	Temp (℃)	t (min)	Entry	Cross-linking reagent (%)	pH	Temp (℃)	t (min)
1	1	1.73	30	1	9	2	3.73	60	1
2	2	2.73	40	5	10	1	4.73	50	5
3	4	3.73	50	15	11	8	1.73	40	15
4	8	4.73	60	30	12	4	2.73	30	30
5	8	2.73	50	1	13	4	4.73	40	1
6	4	1.73	60	5	14	8	3.73	30	5
7	2	4.73	30	15	15	1	2.73	60	15
8	1	3.73	40	30	16	2	1.73	50	30

Though dialdehyde compounds, e.g. glutaraldehyde and glyoxal, have been extensively used as cross-linking reagent for CS, cellulose and starch, it is truly that the mechanism on these reagents bond to CS is still not fully understood. Considering the halobiosic CS has a structure with two hydroxyl groups and an amino group in one glucosamine ring, and the glyoxal molecule has two carbonyl groups, especially the carbon of carbonyl groups due to located in sp^2 easily hybridized to cause attachment from other three atoms lying in the same plane to yield the bond angles among these three atoms a trigonal coplanar structure, e.g. approximately 120°. Moreover, the carbonyl carbon and carbonyl oxygen are available to present several positive and negative charges, respectively, due to the inductive effect from the electronegative oxygen and the resonance contribution of the second structure. To observe these evidences is thus of interest in relation to understanding of the mechanism for using glyoxal to cross-link with CS.

FTIR spectra for CS fiber before and after cross-linking were presented in Figure 7-27. Observed there has a new peak appeared at 1 110 cm^{-1} for chitosan fiber after cross-linking reaction indicated that glyoxals has reacted with the hydroxyls of the glucosamine rings due to acetalization in good agreement with literature. According to Solomons, dissolution of aldehyde in alcohol can yield hemiacetal by nucleophilic addition of the alcohol to the carbonyl group, and this process has a basic feature on hemiacetal, i.e. its —OH and —OR groups to be attached by the same carbon atom.

Moreover, as known that hemiacetal is of unstable and it is easily further isolated in the presence of catalysts, e. g. gaseous HCl, to cause it reacts with the alcohol again to result in an acetal with two —OR groups attached by the same CH group.

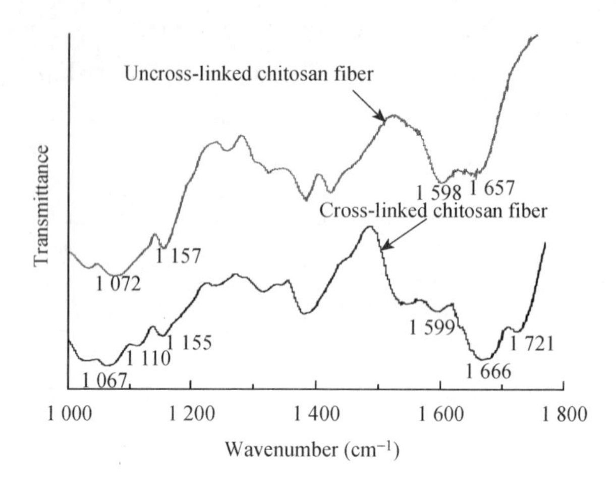

Figure 7-27 FTIR spectra of chitosan fiber before and after cross-linking

Because carbonyl group can also reacted with Schiff base in theory on the base of the presence of amino group due to the addition of a primary amine to an aldehyde in resulting of imines, a Schiff bases to observe this reaction if occurrence between CS fiber and glyoxal is thus another interest. As expected, Figure 7-27 presented a quite strong peak located at 1 666 cm^{-1} assigning to C═N stretching. This is thus available taken as an evidence for supporting expected reaction occurrence. Based on literature, the peaks presented in Figure 7-27 were assigned and summarized in Table 7-14.

Table 7-14 An analysis of chitosan fiber cross-linked with glyoxal based on FTIR spectra presented in Figure 7-27

Groups	Vibration mode	Uncrosslinked CS fiber		Crosslinked CS fiber	
		Peak (cm^{-1})	Intensity	Peak (cm^{-1})	Intensity
O—H	stretching	3 434	very strong	3 434	very strong
N═C—C═O	stretching	—	—	1 721	weak
C═N	stretching	—	—	1 666	very strong
—NH_2	scissor	1 598	general	1 599	general
C—O—C—O—C	asymmetrical stretching	—	—	1 110	weak

Since the primarily orthogonal experiments have showed a considerable reaction condition for applying glyoxal as reagent to cross-link with CS fiber, e. g. the concentration for reagent to be prepared about 2% with pH of 3. 73, and the reaction temperature and time to be about 40 ℃ and 30 min, respectively, to furthermore study these factors how to influence the tenacity in detail is necessary. To do this study, the pH and reaction time were initially confirmed at 3.73 and 30 min, respectively, and to vary the concentration of reagent and temperature. By taking tenacity as a function of various concentrations for reagent, Figure 7-28 showed that the mechanical properties of cross-linked fiber is availably enhanced, however, this seems to be dependent on the concentration of glyoxal because this Figure 7-28 obviously exhibits a peak ignoring the variety of temperature from 25 ℃ to 60 ℃. This is of importance and suggests that glyoxal molecules may play the role as a bridges to interlink CS at its backbone on the carbonyls to cause two or more CS macromolecules joining together to form a network molecule to increase the molecular weight. Moreover, the better concentration for preparation of glyoxal solution based on Figure 7-28 seems to be of about 4%. Of interest this evidence was furthermore observed and certified in Figure 7-29. For example, as the same as plot showed in Figure 7-29 to describe the relationship between the tenacity of CS fiber and the concentration of glyoxal solution, however, corresponding to different reaction times, it was also noted that a better concentration, e.g. of about 4%, is presented. On the basis of Figures 7-28 and 7-29, the effect from the concentration of reagent seems to be clearly. The lower or the higher of the concentration of glyoxal both to be unable for acceptation due to the former unable to lead functional groups to cross-link to CS to form net therefore increasing the tenacity and the latter, however, may cause redundant glyoxal molecules without opportunity to bond on the CS glucosamine rings as expected to increase the mechanical property. Therefore, only when glyoxal concentration is at about 4%, the cross-linking reaction can be accepted to present effective result on the tenacity of fiber. Though this finding seems to be different to primarily orthogonal experiments suggested concentration for glyoxal, e.g. about 2%, the result obtained from Figures 7-28 and 7-29 in reasonable. This is because these two Figures 7-28 and 7-29 consistently indicated that the concentration at 4% is better than that of 2%. In fact, this difference is due to the concentrations designed in orthogonal experiments are 1.73, 2.73, 3.73 and 4.73 (Table

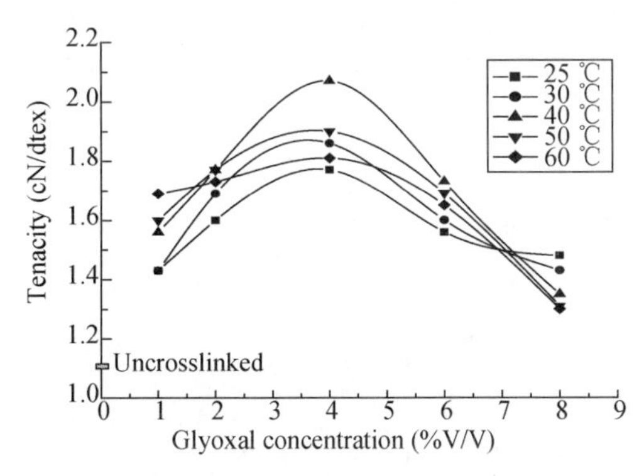

Figure 7-28 Effect of various glyoxal concentration (pH =3.73) on the tenacity of cross-linked CS fiber under different reaction temperatures after reaction of about 30 min

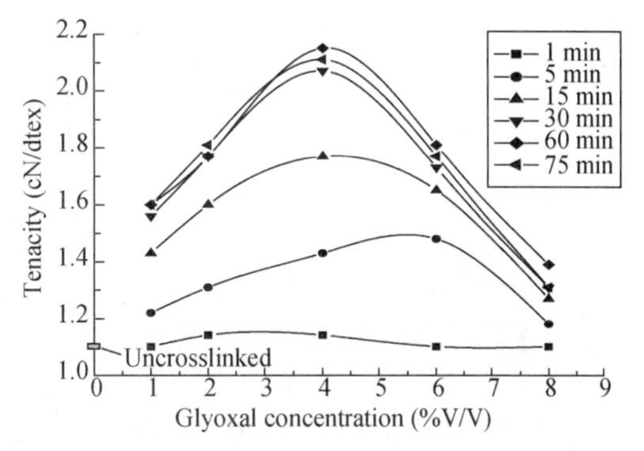

Figure 7-29 Effect of glyoxal concentration (pH =3.73) on the tenacity of cross-linked CS fiber under different reaction times at 40 ℃

7-13), while in Figures 7-28 and 29 the concentration range was extended, e. g. from 1.0% to 8.0%.

From Figures 7-28 and 7-29, it was also noticed that the temperature and time for this case to be better at about 40 ℃ and 60～70 min. Obviously, the temperature resulted in Figure 7-28 seems to be as the same as that of the orthogonal experiments, however, the time seems to be exactly higher comparing to orthogonal experiments. It means that a further study of these factors how to influence the cross-linking reaction is required. Since pH is a considerable factor in relation to preparation of the glyoxal solution, its effect was subsequently studied and plotted in Figures 7-30 and 7-31. Observed, the tenacity of cross-linked chitosan fiber is lower in either lower or higher pH environment, and stronger seems to be corresponding to a pH at of about 4. This suggests that the cross-linking reaction occurred in an acidic environment, e. g. the pH to be in the range of 3～5, is benefit for enhancement of the mechanical properties for CS fiber. The reason for glyoxal cross-linking to CS fiber better in this pH range is perhaps due to Schiff base reaction usually taking place as a nucleophilic addition and that might be susceptibly in the presence of acid catalysis. This can be understood because the primary amine located at the backbone of CS is nucleophilic owing to its unshared pair of electrons on nitrogen. Therefore, once the acidity of the reaction medium increased, the amine could be soon protonated to present non-nucleophilic property to cause the carbonyl group lost chance to take place of addition reaction to yield the low tenacity for CS fiber as Figures 7-30 showed. From these Figures 7-30 and 7-31, the same phenomena as that of Figures 7-28 and 29 were observed again indicating that the temperature and time for cross-linking reaction to be better at 40 ℃ and 60～70 min. In general, these values are in reasonable because a high temperature usually causes degradation taking place for polymer material especially CS. Whereas the reaction time up to about 60～70 min to be better is probably dependence of the amount of CS fiber.

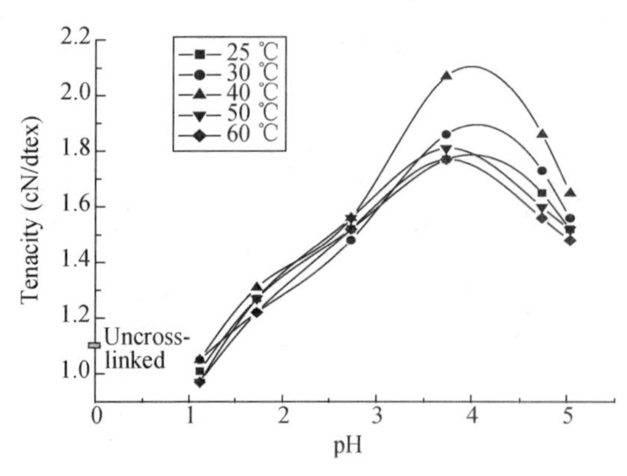

Figure 7-30 **Effect of the variety of pH for glyoxal solution on the tenacity of cross-linked CS fiber under a reaction condition at 40 ℃, glyoxal at 4% and reaction time at 30 min**

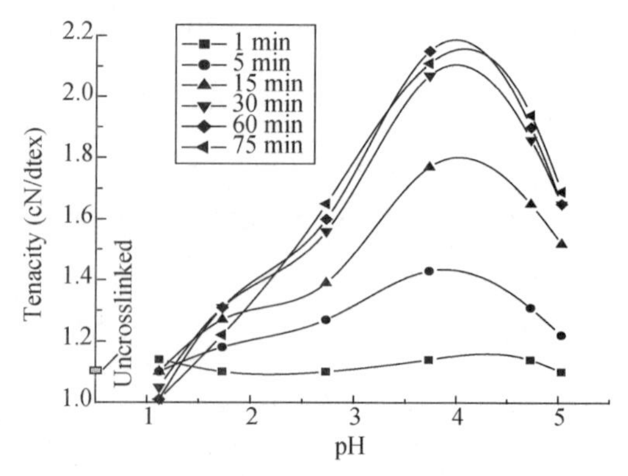

Figure 7-31 **Effect of the variety of pH for glyoxal solution on the tenacity of cross-linked CS fiber under a reaction condition at 40 ℃, glyoxal at 4% and reaction time at 30 min**

On the understanding of the reaction time, it is reasonably because the increase of the mechanical properties for CS fiber using cross-linking method needs to consume time to for stabilization.

The effect of cross-linking on the degree of swelling for CS fiber was showed in Figure 7-32. Observe that the increase of the glyoxal concentration is visible to decrease the swelling degree indicating that the cross-linking reaction is exactly taken placed. Since Figure 7-32 showed swelling behavior has been found in good agreement with Uragami and Takigawa, it is evidently that this case adopted method for description of the cross-linking is capable.

According to Figure 7-32, it is therefore confirmed that the cross-linking process is capable for enhancement of the tenacity for CS fiber. Additionally, this presented swelling behavior suggests that the use of glyoxal as a reagent to cross-link CS fiber may follow a regulation such as: $S = 79.24 - 5.05C + 0.24C^2$. Of which, S and C represent the swelling degree and the concentration of glyoxal, respectively.

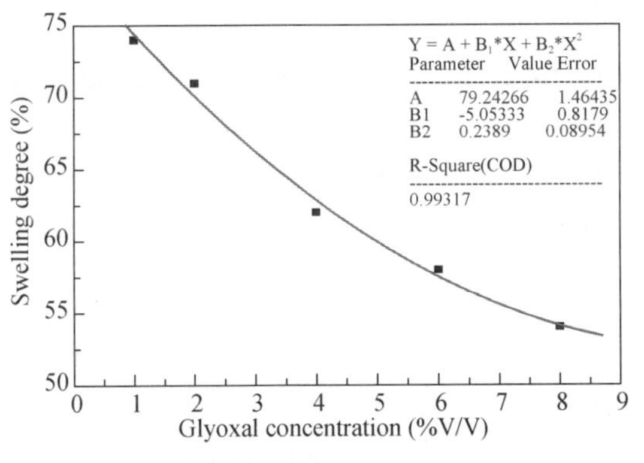

Figure 7-32 Influence of glyoxal concentration on the swelling degree of cross-linked CS fiber

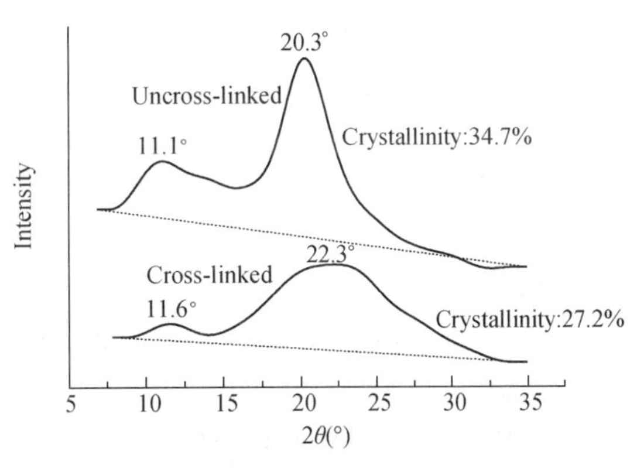

Figure 7-33 XRD patterns of CS fiber before and after cross-linking

Because CS has two crystal types, e.g. α and β, and both belong to the monoclinic system resulting molecular chain not only in regular, rigid and polar, but also easily crystallization, the structure of CS is thus usually presented with different cell parameters. Considering the cross-linking reaction should be a factor to influence the structure for CS fiber, a comparison of the X-ray diffraction spectra for those CS fibers before and after cross-linking with glyoxal was thus presented in Figure 7-33. Clearly, two intense diffraction peaks were visibly located at 11.1° and 20.3°, respectively, for original sample to indicate that these are typical characteristics for CS fiber. Moreover, it is further known that these two intense peaks in relation to the α crystal of CS. Taking this in mind to characterize the cross-linked CS fiber (Figure 7-33), hence these two peaks visibly reduced or shifted for cross-linked CS fiber would be considered that the crystal structure of cross-linked CS fiber might be changed from α to β type. Obviously, this finding is important for understanding the mechanism of cross-linking for CS fiber. Because the β type crystal structure for CS fiber may cause its thermal stability decreasing due to such crystal molecules comprise more amorphous structure than that of α crystal molecules, and the acting forces among β crystal molecules weaker than that of α crystal molecules, it was re-considered that this case found crystal behavior for cross-linked CS fiber seems to be a good explanation for supporting above conclusion that the cross-linking process with glyoxal caused the occurrences of the Schiff base reaction and acetalization.

The crystallinity of the CS fiber was estimated based on literature (*Mo* et al., 1993) suggested Equation (7-1).

$$f_c = \int_{x1}^{x2} I_c(x)\mathrm{d}x \Big/ \int_{x1}^{x2} I(x)\mathrm{d}x \tag{7-1}$$

where: f_c——the crystallinity of CS fiber;

$I(x)$——the gross diffraction intensity of CS fiber;

$I_c(x)$——the diffraction intensity from crystal part of CS fiber; and

x——2θ value.

Based on Equation (7-1), the crystal structure change for CS fiber was known quantitatively, i.e. the cross-linking seems to reduce the crystal component for this fiber from 34.7% to 27.2%. In fact, the formation of the imperfect β-crystal was ascribed to the deterioration of the crystallization for CS after cross-linking in good agreement with literature.

Since the appearance of β-crystal for CS fiber should be to influence its thermal properties, DSC characterization was performed and the related curves were showed in Figure 7-34. Of which, a dehydration phenomenon was initially observed for two samples, e.g. at about 75 ℃ for uncross-linked and at about 78 ℃ for cross-linked CS fiber, respectively. The reason for explaining of these phenomena is probably due to the hydroxyls of CS fiber resulted of hydrogen bonds in connection with the molecules of water during the cross-linking process. In fact, with respect to Figure 7-34 showed a shift, e.g. 3 ℃, for cross-linked CS fiber that might be described due to the moisture evaporation in comparison with uncross-linked sample. It was considered that the cross-linking reaction may embed moisture to result in network for CS fiber.

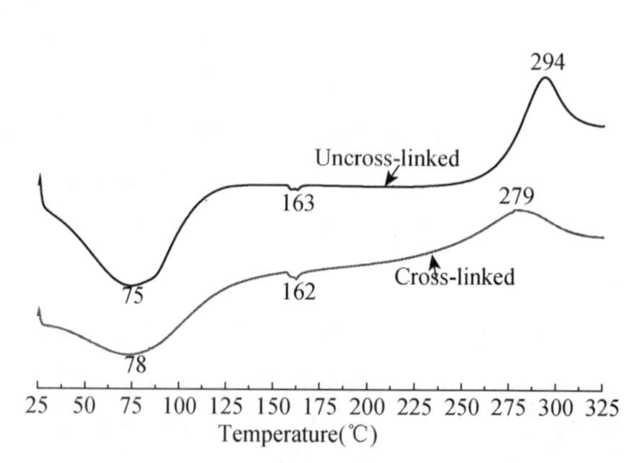

Figure 7-34 DSC curves of CS fibers before and after cross-linking

Because Figure 7-34 also presented two small endothermic peaks at 162 ℃ or 163 ℃, respectively, for both uncross and cross-linked CS fibers, and both these two peaks probably characterized as the glass transition temperature, T_g, it suggests that this thermal property for CS fiber without influenced by cross-linking reaction. Though these T_g data are great different than that of other people reported values, it is, however, supported by *Ahn* et al. Suppose this assignment is correct, the thermal behavior presented in Figure 7-34 is thus considered might be an indication for understanding of the CS fiber after cross-linking. Additionally, the influence of the β-crystal structure occurred for CS fiber on its thermal property was observed in Figure 7-34. This is because the decompose temperature for CS fiber has been found to be 294 ℃ for uncross-linked fiber and about 279 ℃ for cross-linked fiber and both in accordance with literature.

Since above used two methodologies resulted of two conclusions, a hot stage polarizing microscope was further applied to investigate the β-crystal resulted for CS fiber after cross-linking how to influence its thermal property. Relying on CS fibers presented light intensities, and taken temperature variety as a function, a comparison of two samples was showed in Figure 7-35. Of that, the light intensity stronger is for the uncross-linked fiber and the weaker is for the cross-linked fiber.

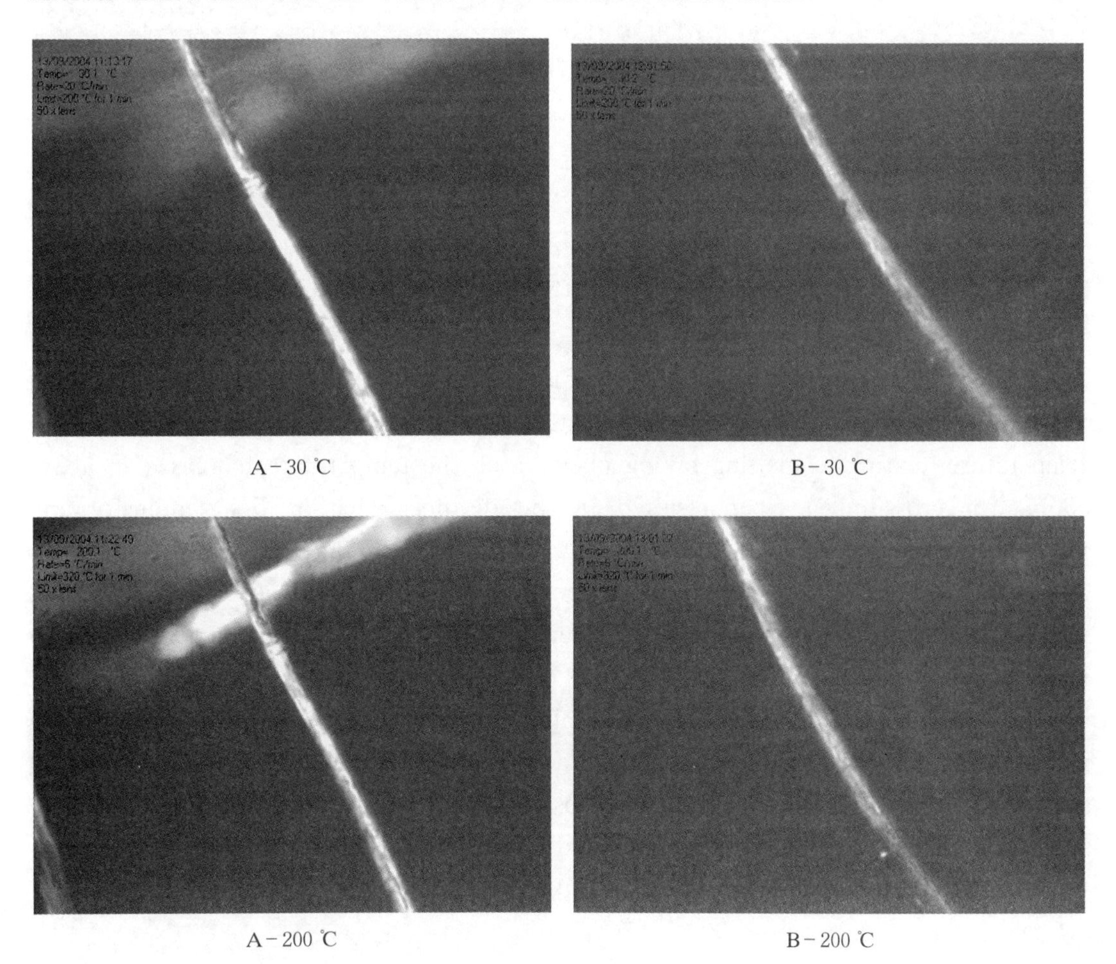

A-30 ℃ B-30 ℃

A-200 ℃ B-200 ℃

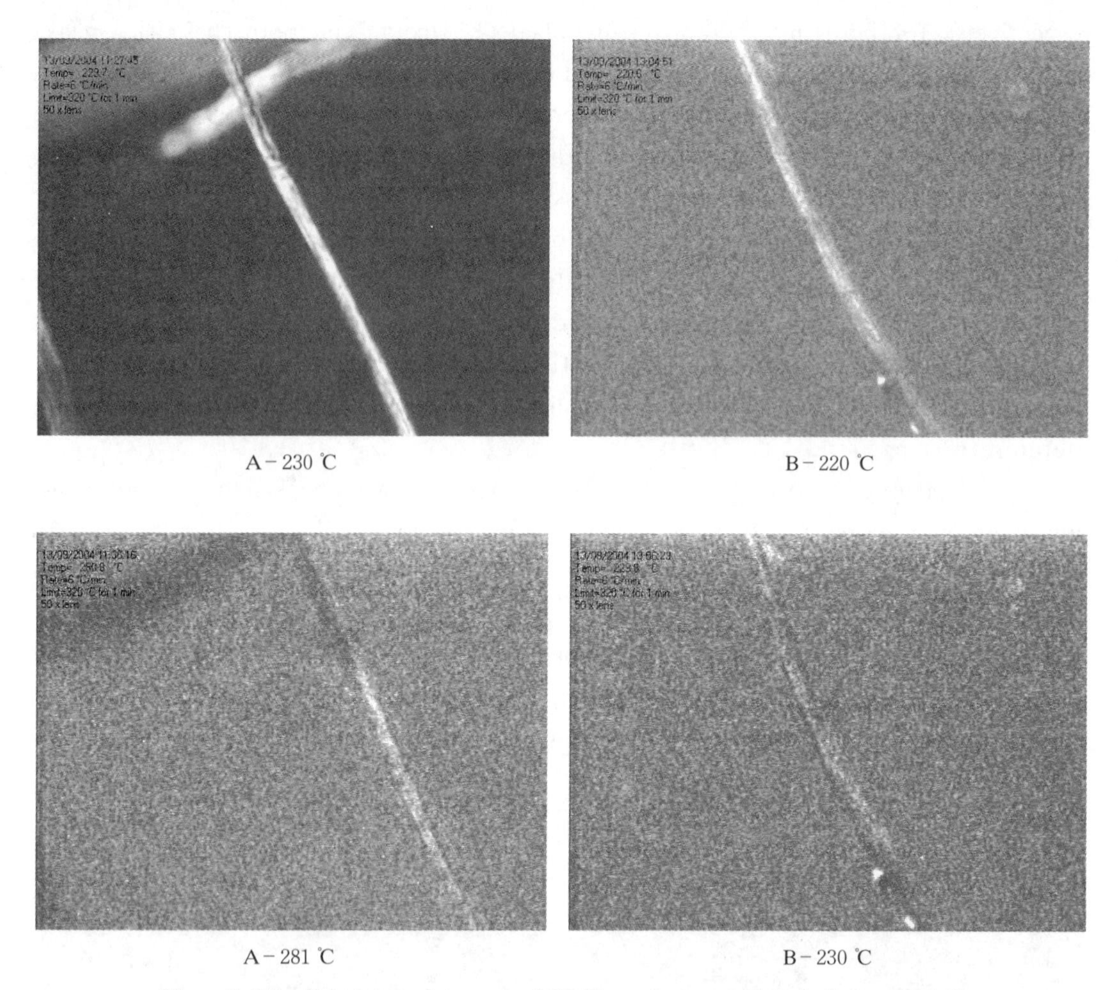

Figure 7－35　Polarizing microscopy of CS fiber, A-uncross-linked, B-cross-linked

Since both these CS fibers have been found to keep their crystal structure when temperature below 200 ℃ and the crystal structure to be damaged at different temperature relating to melting taking place, i. e. the temperature increased to about 230 ℃ for cross-linked fiber resulted in the disappearance of light intensity and meanwhile the uncross-linked fiber still kept its quite integral until the temperature up to about 281 ℃ in resulting the same phenomenon as that of cross-linked sample. Clearly, this indicates that the β-crystal for CS fiber resulted by the cross-linking is indeed to influence its thermal property.

In addition, the influence of cross-linking on the surface of CS fiber was also investigated by SEM. Photographs of CS fiber without cross-linked and cross-linked showed in Figure 7－36 revealed that the cross-linking might be to play a role to coat the surface of CS fiber due to the uncross-linked CS fiber presented roughness with a lot of visible grooves while the surface of cross-linked sample obviously in smooth.

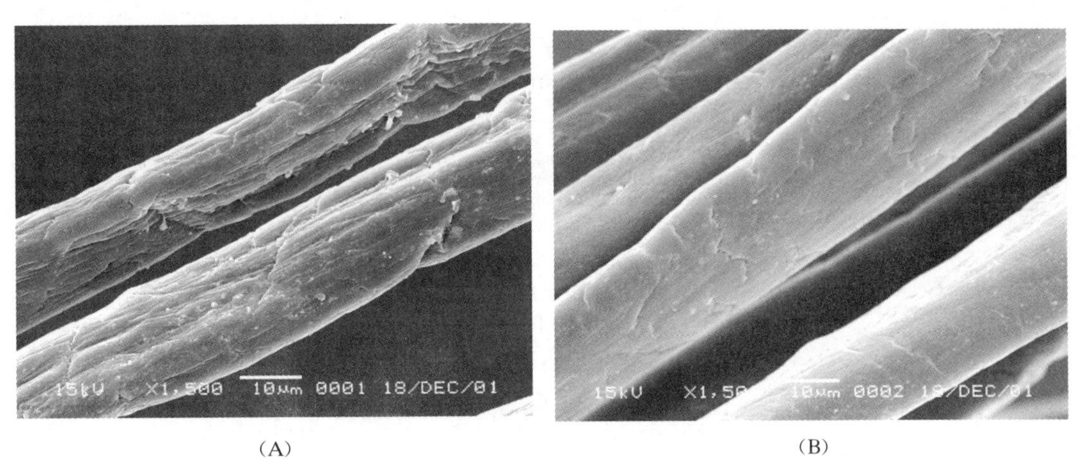

(A) (B)

Figure 7-36 SEM images of CS fiber before (A) and after cross-linking (B)

7.6 Case on formation of cellulose fiber

Since cellulose can be dissolved in a lot of solvents and solutions, to spin cellulose fiber is usually applying the wet process, e.g. wet/solution spinning.

In this case, the cellulose used was a commercial product provided by Shanghai Chemical Fiber Co. and known to have degree of polymerization (DP) of about 500, and α-cellulose content about 90%. Polyoxymethylene (PF) and DMSO were analytical grade solvents purchased from a local chemical company (Shanghai) were used as received without further purification.

Cellulose (10% wt) and DMSO/PF 10% were used to prepare spinning solutions. Solutions were initially stirred and heated to 60 ℃ held for 1 h, then heated to 110 ℃ and held until that temperature transparent. The residual formaldehyde within solution was removed using a vacuum pump at 80 ℃ for several hours. Before spinning, this solution was stirred again and filtered to remove insoluble solid materials.

Laboratory scale wet spinning was employed. Under a pressure of about 30 Kpa and controlled by a metering pump, e.g. 0.2 g/min, spinning solution was allowed to flow into a 100 μm spinnerette die, L/D ratio of 1.0, to form a cellulose filament. Then, this filament was passing and solidified by through two water baths. The first was a coagulation bath with a temperature of about 40~60 ℃; take up rate, V_1, was about 1.2 m/min. The second was a boiling water bath; take up rate, V_2, about 2.4 m/min. Finally, the filament was collected on a winder at a rate of about 10 m/min. After that, the filament was washed with distilled water, then dried in a vacuum oven at 50 ℃ for 24 h. Before analysis and characterization, the filament was held at 20 ℃ and 65% relative humidity.

To apply wet spinning to prepare a filament, the influence from the temperature of coagulation bath was initially investigated. As Table 7-15 reports, we found that both the lowest and highest temperatures are unacceptable for preparation of fiber. The data in Table 7-15 also suggests that the best temperature for coagulation bath seems to be in the range of 40~50 ℃.

Table 7-15 Conditions for wet spinning of cellulose fiber

Temperature of coagulation bath (℃)	Fiber drawing behavior	Fiber breaking intensity (cN/dtex)
20	Unable to draw	—
30	Breaks easily	1.6
40	Good drawing	2.1
50	Good drawing	2.2
60	Breaks easily	1.9
65	Unable to draw	—

Table 7-16 Mechanical properties of cellulose fiber

Fibers	Crystallinity	Crystal grain size (nm)			Orientation	Breaking intensity
	(%)	101	$10\bar{1}$	102	(%)	(cN/dtex)
Cellulose	69.6	1.17	1.66	1.19	70.0	2.20

7.7 Case on directly formation of polyaniline nanofiber by various polymerizations

To form polyaniline, PANI, nanofiber is required due mainly to its electric property. In terms of literature, there have a lot of methods available to spin such nanofibers corresponding to the directly or indirectly, respectively.

7.7.1 Directly formation of PANI nanofiber by solution polymerization

Cyclodextrin, CD, is a family of compounds made up of sugar molecules bound together in a ring and cyclic oligosaccharides, and its typical structure contains six, seven and eight sugar ring molecules in a ring creating a cone shape denoted as the α-, β- and γ-CD, respectively. Due to the great interest on PANI nanofiber, to apply the α-, β- and γ-CD as templates to synthesize PANI nanofibers were performed and compared.

The aniline (99%) and ammonium peroxodisulfate, APS (99%) both obtained from the Sinopharm Chemical Reagent Co., Ltd. located at Shanghai, China. The commercial α-, β- and γ-CDs obtained from Majorbio Biotech Co. Ltd. Shanghai, China without furthermore pretreatment.

A solution polymerization process was used to synthesize PANI nanofibers. The concentration of ANI solution was controlled by taking 0.45 mL ANI to dissolve in

180 mL distilled water to form a very thin solution. This is because a low concentration of ANI solution has been proven available to form PANI nanofibers. An APS aqueous solution was prepared by dissolving 1.09 g APS in 20 mL distilled water, and then rapidly added into the ANI solution within 1 min under a stirring condition, then 1 mmol CD was subsequently add to start the solution polymerization. The total reaction time was controlled at 24 h, at 25 ℃, and obtained dark green PANI sample was washed by deionized water for several times until the filtrate to show colorless then be vacuum oven dried at 60 ℃ for 24 h.

FESEM images of three PANI nanofibers template by α-, β- and γ-CDs were presented in Figure 7-37. It was observed that three samples all showed a mat where the fiber shape is clearly, especially in rigid and uniform. This suggests that the use any kind of those CD as template can fabricate uniform PANI nanofibers. In terms of statistics software, the average fiber diameter of those fibers was estimated and appeared in Figure 7-37. According to Figure 7-37, the diameter of those PANI nanofibers templated by α-CD is in the range of 27～45 nm and of which about 50% is at 34 nm (Figure 7-37, top), by β-CD is in the range of 17～28 nm and of which about 60% is at

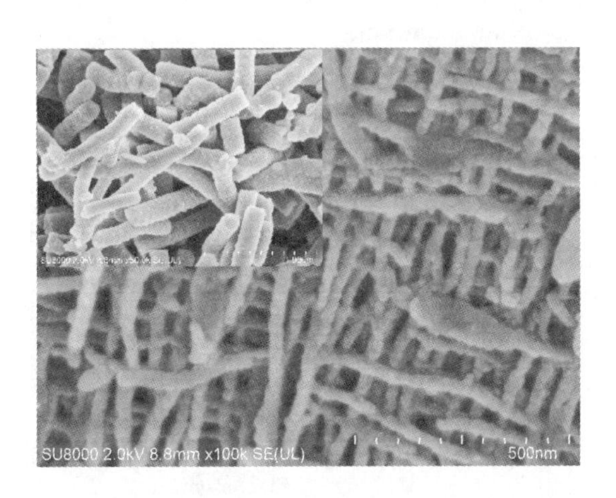

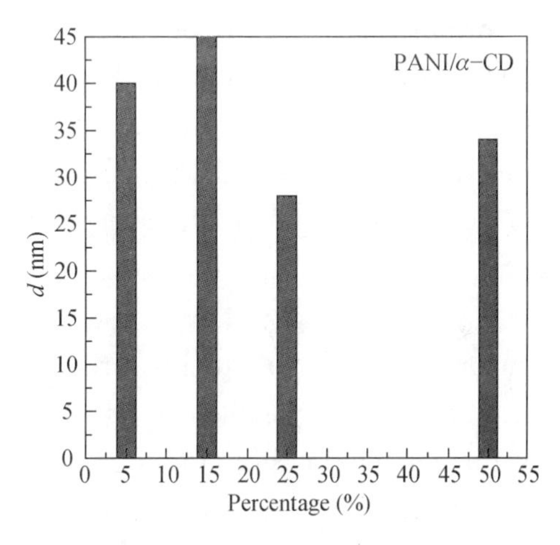

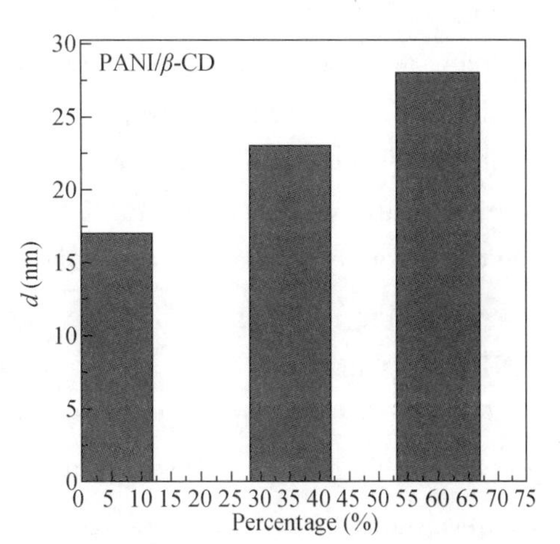

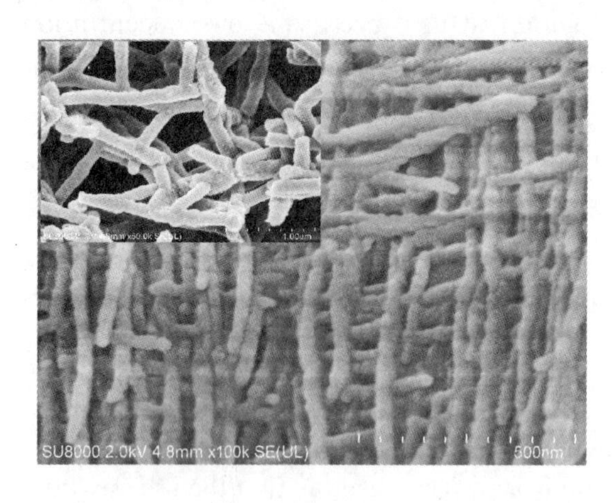

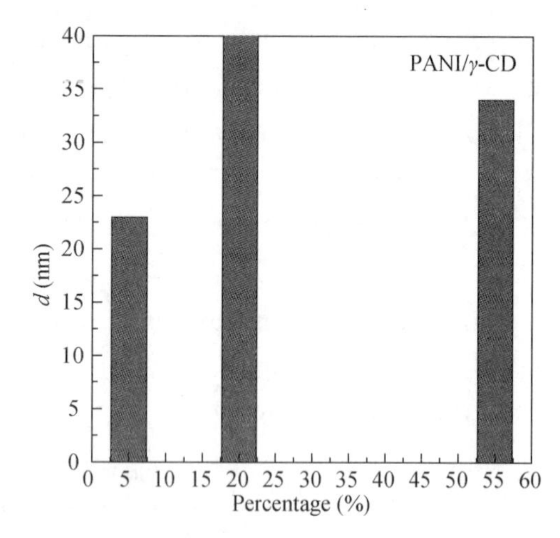

Figure 7-37　FESEM images of three PANI nanofibers templated by α-, β- and γ-CDs, respectively

28 nm (Figure 7-37, middle), and by γ-CD is in the range of 22~40 nm and about 55% is at 32 nm (Figure 7-37, bottom), respectively. These results are interested because all measured diameters are smaller as compared with literature reported values on using CD as template. This suggested that the solution polymerization might be a better method for forming PANI nanofibers than that of the other techniques.

In order to understand the diameter of PANI nanofibers is how to be influenced by used CD, a plot was made and showned in Figure 7-38 by taking the average fiber diameter, d_a, as a function of the Lifshitz-van der Waals, LW, interaction component, γ^{LW}, of those three CDs. Since these two parameters appeared a good linear relationship, this suggests that the diameter of those PANI nanofibers is dominated by the γ^{LW} of CDs. This is possible because the used CD is a dominator in self-assembling PANI nanofiber by the LW interactions to agree with literature. In fact, the PANI and CD both are strongly in the LW component.

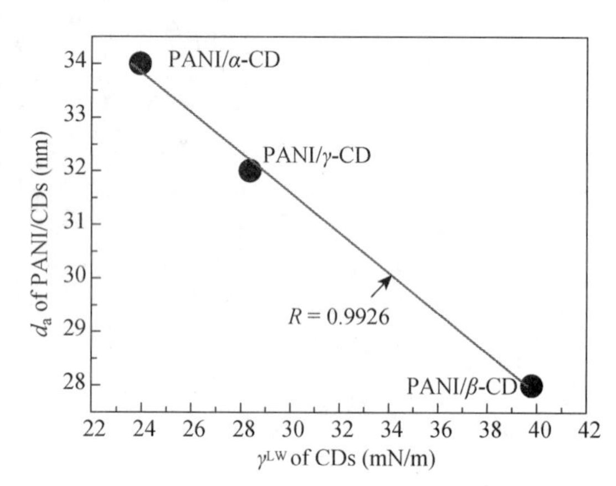

Figure 7-38　Relationship between the average diameter, d_a, of three PANI nanofibers templated by α-, β- and γ-CDs, respectively, and the Lifshitz-van der Waals interaction component, γ^{LW}, of the used CDs

According to Figure 7-38, the use of those three CDs can control the diameter of PANI nanofibers.

FTIR spectra of those three PANI nanofibers were showed in Figure 7-39. These spectra exhibited intense absorption peaks at 1 573 cm^{-1} corresponding to the stretching of quinonoid, at 1 493 cm^{-1} due to the benzenoid ring vibration, at 3 430 cm^{-1} due to the

N—H stretching and at 2 924 cm^{-1} due to the C—H stretching.

The main IR peaks of CDs are located within the ranges of 3 315～3 337 cm^{-1}, 1 642～1 658 cm^{-1}, 1 153 cm^{-1}, 1 024～1 028 cm^{-1}. Since these peaks are overlapped in PANI/CDs (Figure 7-39), this suggested that the PANI and CD formed the host-guest inclusion complex. Since Figure 7-39 presented two peaks at about 997 cm^{-1} and 970 cm^{-1} contributed by CD, it is considered that the use of CD as template enhanced the π-electron delocalization for PANI nanofibers.

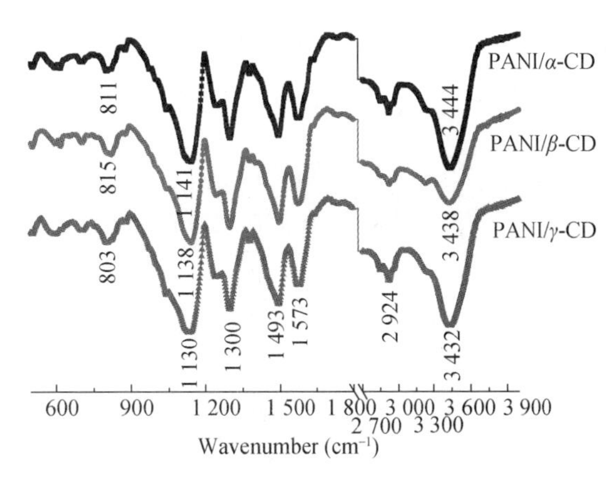

Figure 7-39 FTIR spectra of PANI nanofibers templated by α-, β- and γ-CDs

The XRD patterns of three PANI nanofibers were presented in Figure 7-40. Though there showed two typical 2θ peaks corresponding to the reflection planes of (020) and (200) corresponding to the periodicity parallel and perpendicular to the polymer chains, respectively, it is found that the right peak is stabilized at 25.6° for all three PANI samples and the left peak is shifted from 20.2 to 19.9 then 19.7, respectively, for PANI dominated by α-, β- and γ-CD. This suggested that these peaks were passivated for those PANI nanofibers due to formed different host-guest inclusion complexes due to those different CDs induced different crystallinities. This indicated that the CD is not only a template to guide the nanofiber formation but also a dominator to control the diameter of PANI nanofibers.

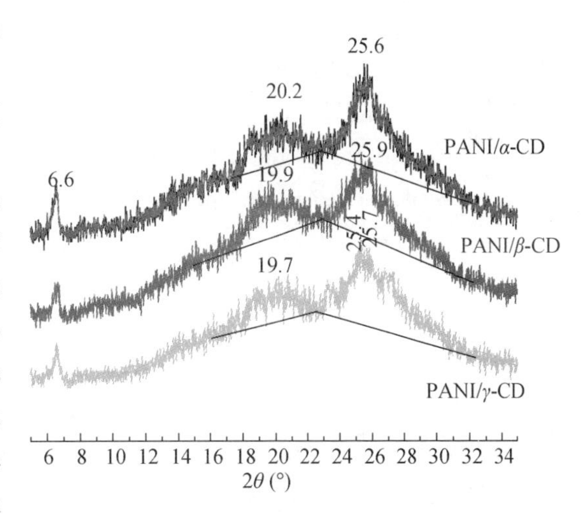

Figure 7-40 XRD patterns of PANI nanofibers templated by α-, β- and γ-CDs

7.7.2 Directly formation of PANI nanofiber by emulsion polymerization

The carbon nanotube (CNT) has been known with excellent Young's modulus, good flexibility, high electrical and high thermal conductivity, and usually recognized as reinforcement broadly applied in high performance and multifunctional composites. Additionally, CNT is also reported to blend with PANI to enhance the conductivity of PANI. Notable, since CNT is insoluble in any solvents and difficult disperse well due to precipitation immediately to form bundles, to do a pretreatment of CNT is required and necessary. Moreover, the pretreatment has been found to destroy the electronic network of CNT and decrease the degree of electron delocalization and activity of CNT to this can be performed by a lot of methods. As has been further found that the exfoliating of bundles and preparation of individual CNTs in solution could utilize polymers, for example, the polynuclear aromatic compounds, surfactants, and some biomolecules, when CNT was functionalized in noncovalent to cause the attached chemical handles unable to affect the electronic network of CNT because the noncovalent interactions would be occurred in CNT via the van der Waals forces or $\pi \rightarrow \pi^*$ stacking and controlled by thermodynamics.

Lignosulfonate (LGS) is a natural polymer with sulfonated groups in water-soluble. This polyanion has been investigated of its surface properties and found it has low surface free energy. Therefore, we applied LGS to pre-treat CNT then furthermore applied modified CNTs to synthesis of PANI nanofibers with soluble and high conductivity properties. Experimentally, a calcium LGS (LsCa) was employed to modify the surface of CNTs via a self-assembly technique. For this step, it is expected to yield CNT-LGS to lead CNTs well dispersed in the solution of anilinium monomers to benefit CNT as a template to allow the anilinium monomers directional aligned on the surface of CNT then presented in polymerized PANI.

Multiwalled carbon nanotubes (MWNTs) with the purity above 90% purchased from Chengdu Institute of Organic Chemistry, Chinese Academy of Sciences were used. As known, these CNTs were produced by the method of chemical vapor deposition and their lengths are ranged from several hundred nanometers to several micrometers and its averaged outer diameter is about 10～30 nm. A calcium LGS with an average molecular weight of 100 000 purchased from Jiangmen Sugar Cane Chemical Factory, Guangdong of China was used as received. Based on the producer, this LGS is mainly composed of phenylpropane segments and sulfuric acid groups, and the lignin component is more than 55%, deoxidized sugar is less than 12%, water-insoluble components are less than 1.5%, and the moisture is less than 9%. The pH of this LGS is measured in a range of 5～6. The aniline, ammonium peroxodisulfate (APS) and analytical grade solvents were used all purchased from Sinopharm Chemical Reagent Co., Ltd. Shanghai, China. Distilled water was used throughout the experiments.

The modification of MWNTs by LGS was started by mixing 0.5 g MWNTs with 5 g LsCa in an agate mortar for 3 h by hand initially. Then a small quantity of water was added to avoid agglomeration. The obtained brownish slurry was added to 200 mL buffer and subsequently sonicated using an ultrasonic tip to stimulate the micelle formation. After that, the mixture was sonicated for about 30 min and centrifuged at 15 000 r/min to remove the residual solution. The obtained MWNTs-LsCa were washed with 300 mL distilled water three times to remove the excess LGS. The obtained MWNTs-LsCa were finally dispersed in water to reach a concentration of 3.0 mg/mL. In this case, the excess LGS was analyzed by element analysis of MWNT before and after LGS modification. The obtained MWNTs-LsCa was furthermore applied as a template to synthesis PANI/MWNTs-LsCa. By dispersing 0.465 g MWNTs-LsCa in 50 mL HCl solution (1 mol) under the sonication and room temperature condition, 25 ℃, for 48 h, 4.65 mL aniline monomer (0.05 mol) was added to the suspension under a stirring condition for 30 min when a brownish black emulsion was yielded. Meanwhile 11.41 g APS (0.05 mol) was dissolved in 50 mL HCl solution (1 mol) then slowly dropwise added to aforementioned stirred reaction mixture to keep the stirring for another 30 min. In this step, it was observed that the suspension color was gradually changed after few minutes when the dark color changed to the green indicating the PANI polymerization approaching to the end. The whole polymerization process was carried out at 0 ℃ under nitrogen environment and stirring condition for about 24 h. The obtained PANI/MWNTs-LsCa composite was filtered and washed with 300 mL deionized water for four times, then dried under a vacuum condition, 50 ℃, for about 72 h. A pure PANI was also prepared as the same as above and taken as a reference.

The wide angle X-ray diffractograms of PANI/MWNTs-LsCa and pure PANI were presented and compared in Figure 7-41, where the 2θ scan showed typical diffraction patterns of PANI, that is, two peaks observed at around $2\theta = 19.69^\circ$ and 25.07° represent a reflection plane of (020) and (200), respectively. The 2θ peak at 19.69° may ascribe to the periodicity parallel to the polymer chain, and the peak at 25.07° may be the periodicity perpendicular to the polymer chain. The XRD pattern of MWNTs-LsCa doped PANI shown two peaks at 20.20 and 25.04° similar to the pure PANI, whereas its characteristic peak located at 19.69° representing the crystalline has been found shifted to 20.20°. This means that the narrower intervals are existed among the lattices, and this leads to form denser structures in novel PANI. In contrast to the pure PANI, the new peak appeared at 15.21° in PANI/MWNTs-LsCa represents a reflection plane (011). In Figure 7-41, the PANI/MWNTs-LsCa clearly revealed two new weak peaks appeared at 35.53° and 42.78°, respectively, that may be an indication of the formation of highly ordered structure in PANI/MWNTs-LsCa. Table 7-17 reported and

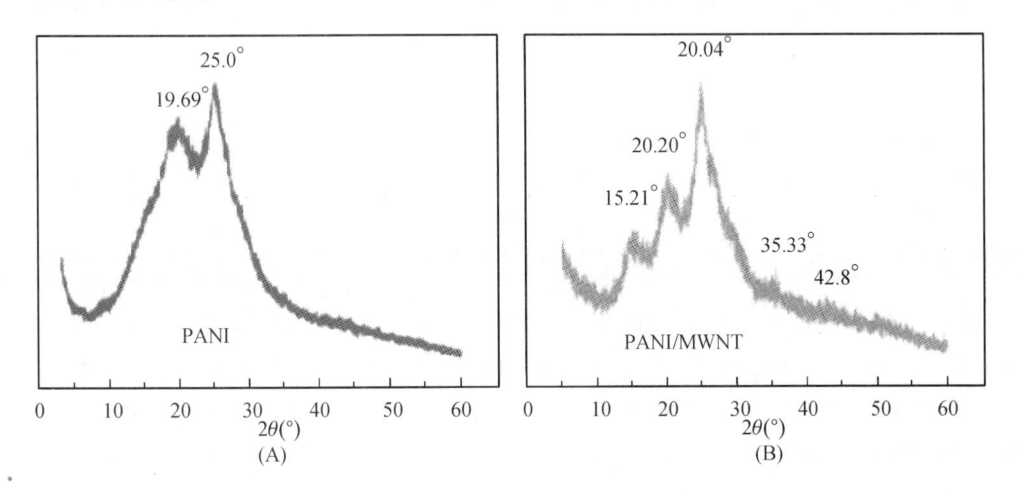

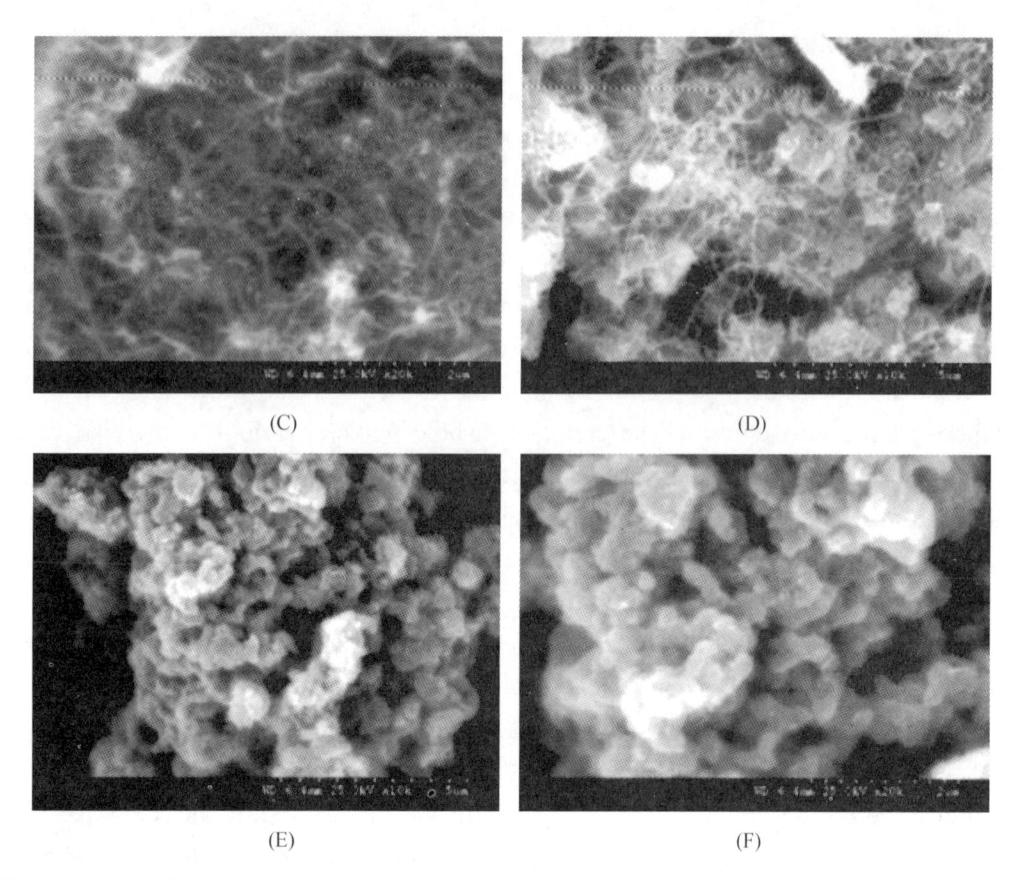

Figure 7-41 FESEM images and XRD patterns of PANI nanofibers template by LGS-modified MWCNT.

compared the crystallinity of PANI/MWNTs-LsCa and referenced PANI, which indicated that the space orientation occurred in PANI/MWNTs-LsCa due to the addition of MWNTs-LsCa caused PANI chains planar alignment. Therefore, both the PANI chains and the MWNTs-LsCa alternatively arranged each other under the electrostatic force, and the electrical conductivity of PANI/MWNTs-LsCa increased greatly with the improvement of the crystallinity.

Table 7-17 The elemental analysis of MWNTs, MWNTs-LsCa, PANI and PANI/MWNTs-LsCa

Samples	C (%)	H (%)	N (%)	S (%)	C/N (mol/mol)	C/H (mol/mol)
MWNTs	97.11	0.77	$\leq$0.05	$\leq$0.05	—	10.51
MWNTs-LsCa	88.21	2.30	0.14	2.32	—	3.20
PANI	53.74	5.73	10.32	$\leq$0.05	6.08	0.78
PANI/MWNTs-LsCa	59.39	4.99	7.82	1.43	8.86	0.99

The FESEM images of the MWNTs, MWNTs-LsCa, and PANI/MWNTs-LsCa were associated shown in Figure 7-41. Figure 7-41(C) shows a typical SEM image of the pristine MWNTs, which are highly tangled with each other, ropes with a smooth surface and have rarely visible ends. In Figure 7-41(D), the SEM image of MWNTs-LsCa exhibit a rough surface suggesting the surface of the pristine MWNTs modified by LsCa in accordance with the conclusion of FT-IR and elemental analyses in below. The PANI thin layer deposited on the surface of the MWNTs-LsCa is observed in Figure 7-41(E, F), and it also revealed the PANI wrapping the MWNTs-LsCa after composite formation in accordance with the polymerization mechanism as aforementioned. In addition, Figure 7-41 showed that more than 90% samples consist of nanofibers with the diameter less than 250 nm, and most fibers are bent or tangled indicating high flexibility. This is of interest and may mean that the interactions occurred between the PANI chains and MWNTs-LsCa has overcome the van der Waals force. In other words, the nanofibers like novel PANI means that the CNTs were coated by PANI because the surface of CNT is easily linked after LGS modification. This is possible because a tubular layer of uniform coated PANI has been observed clearly on the surface of the MWNTs-LsCa. This fact is of interest for preparing PANI because the observed phenomena in Figure 7-41 indicated that the polymerization of aniline inside the MWNTs-LsCa is hindered by the restricted LsCa.

Table 7-18 summarized and compared the XRD results of PANI nanofibers.

Table 7-18 Profiles of parameters from resolved WAXD pattern of PANI and PANI/MWNTs-LsCa

Samples	2θ position	Full width at half maximum	Crystallinity (%)
PANI	19.69°	0.734	32.62
	25.07°	0.968	
PANI/MWNTs-LsCa	15.21°	0.405	48.27
	20.20°	0.545	
	25.04°	0.680	
	35.53°	0.394	
	42.78°	0.278	

The use of MWNTs-LsCa as a template to synthesis PANI/MWNTs-LsCa was described by recorded FTIR spectra and related photos as combined to show in Figure 7-42, where the pristine MWNTs deposited at the bottle bottom in agglomeration similarly as literature supporting the hydrophilicity of MWNTs were greatly improved by LsCa. When the aniline monomers added to the solution of MWNTs-LsCa, the color was gradually changed to brownish black to form an emulsion. This phenomenon indicates that the interactions between the anilinium monomers and the hydrophilic groups, SO_3^-, of the MWNTs-LsCa were taken place. In Figure 7-42, the absorption of anilinium monomers on the surface of MWNTs-LsCa was extremely because an increase of the confusion degree on this system and an enhancement of the stability on emulsion are obviously. Adding APS to the emulsion, the polymerization was prolonged along at the surface of MWNTs-LsCa to allow them wrapping with PANI chains to lead PANI/

MWNTs-LsCa in a higher yield, for example, 72%, as comparing to the referenced pure PANI, for example, 53%. The increased yield indicates that the MWNTs-LsCa was indeed incorporated in PANI. This was also proven by the elemental analysis (Table 7-17), because the novel PANI is greater in the C and S elements and its C/N ratio as comparing to the pure PANI.

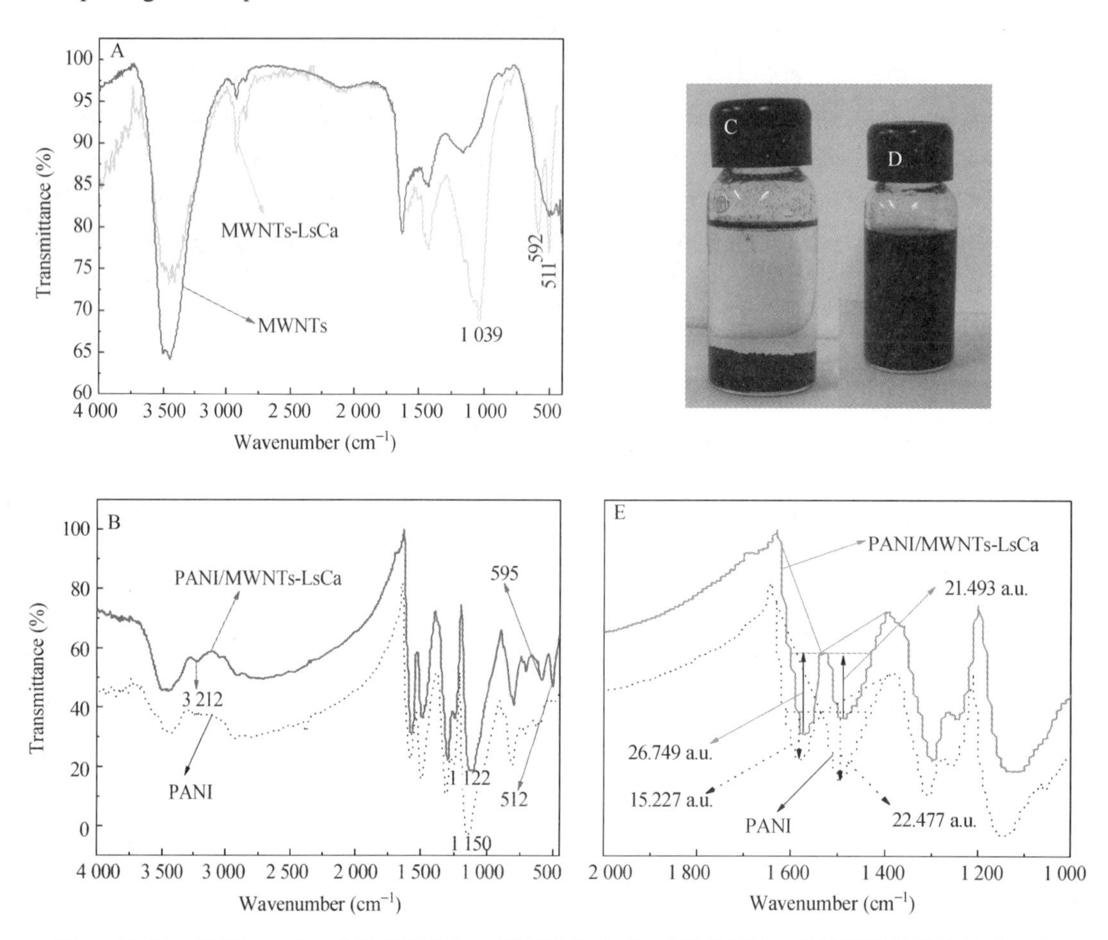

Figure 7-42 FTIR spectra. (A) MWNTs and MWNTs-LsCa, (B) PANI and PANI/MWNTs-LsCa; the quinoid to benzenoid intensity ratio of (E) PANI and PANI/MWNTs-LsCa; Photographs of aqueous suspensions of (C) MWNTs, and (D) MWNTs-LsCa after adding aniline for 2 h

In Figure 7-42, the FT-IR spectra of MWNTs, MWNTs-LsCa, PANI/MWNTs-LsCa, and pure PANI were recorded in the region of 4 000~400 cm^{-1} where the band at 3 450 cm^{-1} is attributed to the O—H stretching contributed by both MWNTs and MWNTs-LsCa contained water molecules. The band at 2 920 cm^{-1} is attributable to the CAH stretching and has been observed for both MWNTs-LsCa and MWNTs. Since the presented intensity is stronger for the former and weaker for the latter, this means that the MWNTs were modified by LsCa to support the element analysis results (Table 7-17). The peak located at 1 630 cm^{-1} is typically for MWNTs due to the —C=C— stretching. In Figure 7-42, a comparison of the C—O stretching at 1 039 cm^{-1} for

MWNTs and MWNTs-LsCa suggested that this peak could be employed to prove the presence of LsCa in MWNTs because of it from LGS. The peak located at 592 cm^{-1} and 511 cm^{-1}, respectively, appeared in MWNTs-LsCa are contributed by the SO_3^- groups of LsCa and the SO_2 scissoring model. By comparison of the FT-IR spectra for PANI/MWNTs-LsCa and PANI (Figure 7-42, B), it was found that the quinoid and benzenoid ring vibrations are presented at 1 580 cm^{-1} and 1 480 cm^{-1}, respectively, and the intensity for former is higher as compared with the latter suggesting the presence of PANI in its ES form. In Figure 7-42, the band at 3 440 cm^{-1} is attributed to the NAH stretching. The band at 1 300 cm^{-1} was attributed to the C—N stretching of benzenoid unit, and the 800 cm^{-1} peak was corresponding to the 1, 4-disubstituted aromatics in PANI. The O—H stretching has been found to appear at 3 212 cm^{-1} in PANI/MWNTs-LsCa to prove the presence of LsCa, and the peaks located at 595 cm^{-1} and 512 cm^{-1} proved the presence of SO_3^- groups in novel PANI. In Figure 7-42, the PANI/MWNTs-LsCa presented an increased intensity ratio, 1 580/1 480 cm^{-1}, relating to the quinoid (Q) to benzenoid (B) which is about 1.25 higher than that of the pure PANI, for example, 0.68. This suggests that the novel PANI with quinoid units due to the π-stacking interactions taken placed. Obviously, these interactions are available to promote the stabilization of the quinoid ring structure in the PANI. In terms of Figure 7-42, several peaks in the novel PANI have been also increased of the intensities or wavenumber shifted to the peaks of referenced PANI. For example, the 1 150 cm^{-1} peak was shifted to 1 122 cm^{-1} (Figure 7-42, B) and it has been designated as "lectronic-like absorption" peak (N=Q=N) in good agreement with our conductivity measurements. This indicates that the interactions occurred between the PANI chains and MWNTs-LsCa has caused the increase of the effective degree of electron delocalization and enhancement of the conductivity of PANI. Therefore, this proven that the role of MWNTs-LsCa played in this case is a chemical dopant for PANI.

The UV-visible spectra of PANI and PANI/MWNTs-LsCa were showed and compared in Figure 7-43. Observe three main peaks were observed located at 338 nm, 606 nm, and 684 nm, respectively, and they could be assigned to the $\pi \rightarrow \pi^*$ transition of benzoid ring, the excision transition from higher energy occupied benzenoid moiety to lower energy unoccupied quinoid moiety and the polaron transitions, respectively. Since the pure PANI showed a new

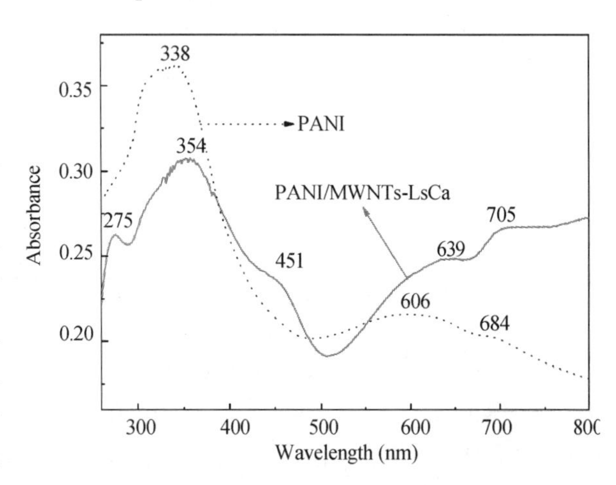

Figure 7-43 UV-vis spectra of PANI and PANI/MWNTs-LsCa

peak located at 275 nm attributed to the $\pi \rightarrow \pi^*$ transition of MWNTs in agreement with literature, and a peak at around 451 nm is due to the electronic state induced by the doping of the sulfonic acid groups of MWNTs into the PANI, these indicated that the interactions occurred between the MWNTs-LsCa and PANI that facilitated the electron delocalization and enhanced the conductivity of the novel PANI. The peaks at 275 nm and 451 nm along with the peaks of PANI indicated the doping of CNT in PANI exactly. In the UV-visible spectrum of PANI/MWNTs-LsCa, the peak at 684 nm assigned to the site selective interactions between MWNTs-LsCa and quinoid ring of PANI chains has been found shifted to 705 nm, this suggests the facilitating charge transfer from quinoid unit of PANI chains to MWNTs-LsCa.

Because LsCa has the low surface energy, the use of it caused modified MWNTs thus probably played as a template to synthesis PANI as expected to enhance the solubility for yielded PANI. To check this assumption, several organic solvents were employed as probes as Table 7-19 described to measure the solubility of obtained novel PANI. As Table 7-19 indicated, the pure PANI can be dissolved only in NMP, DMSO, and DMF, whereas the novel PANI presented good solubility in a lot of used solvents, that is, the good solvents for PANI/MWNTs-LsCa are found to be the NMP, DMF, DMSO, and methyl ethyl ketone (MEK) especially NMP and DMSO. According to Table 7-18 presented parameters, it is probably assumed that the solubility of PANI is influenced by the dielectric constant and solubility parameter of these solvents, and the obtained novel PANI has a modified structure than that of the pure sample thus enhanced of the solubility.

Table 7-19 The solubility of the PANI/MWNTs-LsCa in different solvents

Solvents	Dielectric constant	Solubility parameter	Solubility
NMP	32.2	23.1	+ + +
DMSO	47.2	24.6	+ + +
DMF	38.2	24.8	+ + +
methyl ethyl ketone	18.6	19	+ +
chloroform	4.8	19	+
butylacetate	5.1	17.4	+
xylene	2.4	18	×

×, +, + + and + + + represent solubility (g/100 mL): <2, <3, <6 and <9, respectively.

A relationship between the solubility of novel PANI with the dielectric constant of solvents was plotted in Figure 7-44 for understanding the solubility of this case obtained novel PANI. In contrast to the pure PANI, the solubility increased visible for novel PANI is reasonable due to the incorporated LsCa modified CNTs contained strong polarity group, SO_3^-. The solubility of novel PANI was also described by measuring the conductivity by dissolving the PANI in different solvents (Table 7-18).

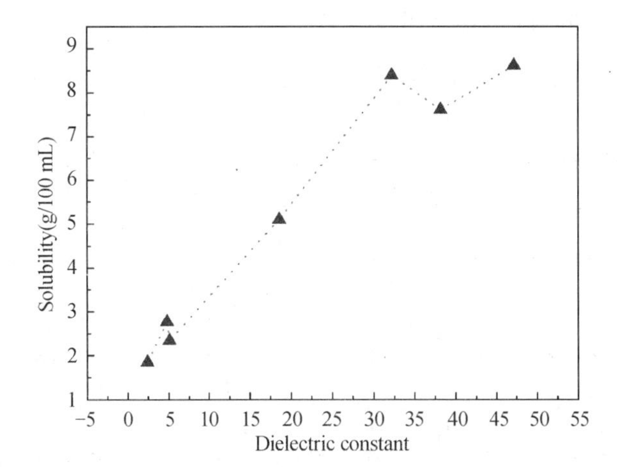

Figure 7-44 The conductivity vs. concentration of PANI/MWNTs-LsCa in different solvents

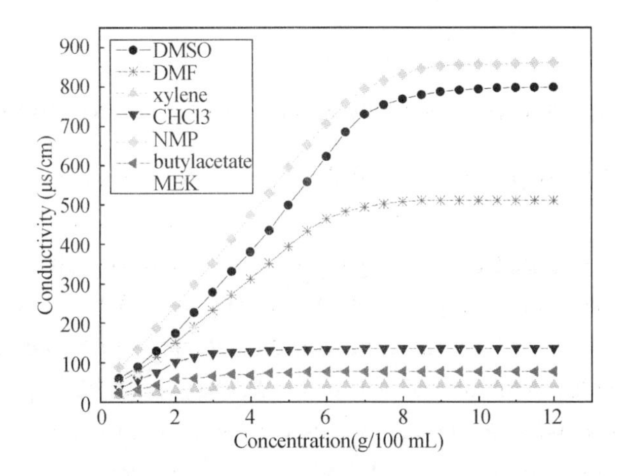

Figure 7-45 The relationship between dielectric constant and solubility for PANI/MWNTs-LsCa

Figure 7-45 presented that the increase of the concentration of solvents, the increased is the conductivity was of PANI/MWNTs-LsCa. Since the conductivity is proportional to the solubility, the curve presented in Figure 7-45 is a support of above conclusion. However, Figure 7-45 revealed information is importantly for understanding the solubility of the novel PANI because it reflected a relationship between the solvent and related concentration that is useful for application. In addition to the measurement of the liquid conductivity as Figure 7-45 described, the solid electrical conductivities of pure and composite PANI were measured according to the standard four point probe technique, and found the electrical conductivity of pure PANI prepared without MWNTs-LsCa is 0.20 S/cm which is agreed with literature, and at the room temperature the conductivity of the PANI/MWNTs-LsCa composite is 5.43 S/cm greatly about two order of the magnitude than that of the pure PANI. The conductivity enhanced about 279 times for this case obtained novel PANI as compared with the referenced PANI sample suggested that the doping is effectively as expected where the charge transfer from the quinoid unit of PANI to the CNTs.

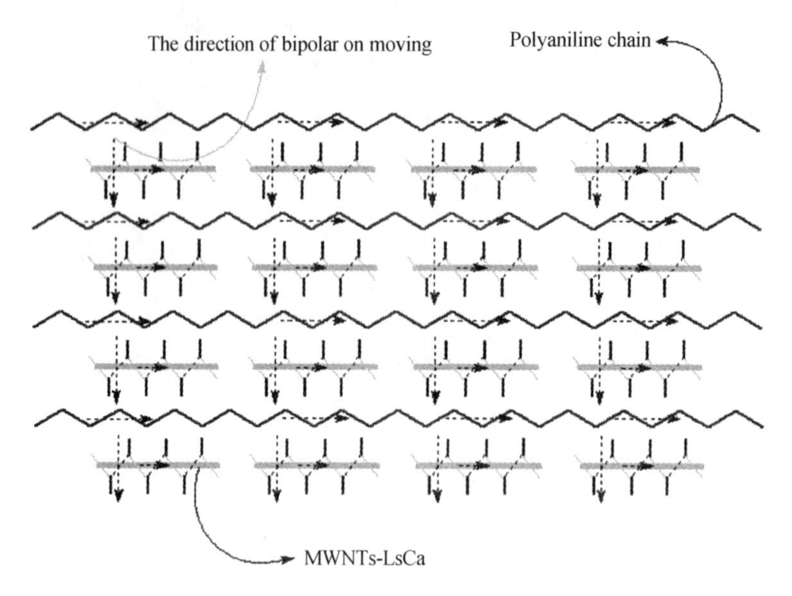

Figure 7-46 The schematic for the high conductivity of PANI/MWNTs-LsCa

Since Figure 7-42(D) presented well dispersion on MWNTs-LsCa in the solution of anilinium monomers, and the reason is because the surface of MWNTs contains the hydrophilic group, SO_3^-, that is benefitable the self-assembly. Therefore, the anilinium monomers can be directionally aligned on the surface of MWNTs-LsCa. When MWNTs-LsCa added into APS, the polymerization would be started along the surface of MWNTs-LsCa, and the molecular structure of the novel PANI would be depicted. This suggests that the role of MWNTs-LsCa played in this case is a dopant and a crosslinker. Since CNT is a good electron acceptor, and PANI can be considered a good electron donor, the large aspect ratio and surface area of LGS-modified CNTs, for example, MWNTs-LsCa, in this case may serve as "conducting linker" to link PANI chains to increase the effective percolation as confirmed by FT-IR (Figure 7-42) and UV-vis spectra (Figure 7-43). Hence, the bipolaron of the PANI chains can transfer through MWNTs-LsCa and other PANI chains as Figure 7-46 described, where the dash line arrow represents the direction of bipolaron moving and the degree of electron delocalization increased visible to cause the enhancement of the conductivity of PANI/MWNTs-LsCa.

7.7.3 Directly formation of PANI nanofiber by electric polymerization

Fabrication of polyaniline, PANI, nanostructure is of importance for application and has led a lot of methods developing. According to literature, PANI nanostructures could be prepared using the emulsion polymerization, solution polymerization, surfactant-assisted synthesis, electrochemical method, interfacial polymerization, seed polymerization, template or template-free method and electro-synthesis. Among these methods, the electro-synthesis showed advantages in simple preparation procedure, accurate control of the initiation and termination steps, and formed purer PANI than that from chemical methods due to the absence of additional species.

Up to now, the electro-synthesis of PANI has been performed with various forms such as the constant potential (potentiostatic), constant current (galvanostatic), cyclic voltammetry, pulse constant potential (pulse potentiostatic), pulse current (pulse galvanostatic), and step-wise galvanostatic methods.

We have developed a novel electro-synthesis method to form PANI nanofibers. Experimentally, an electrostatic generator was employed to assist the normal solution polymerization by taking the positive and negative electrodes alternatively linked to either the metal reactor or the ANI solution to form two different electric cycles corresponding to the electrostatic interactions, EI, reduce (+/−) or enhance (+/+, or −/−). An electric-free sample was prepared as reference.

The aniline (99%), ammonium peroxodisulfate, APS (99%) and citric acid were obtained from the Sinopharm Chemical Reagent Co., Ltd. located at Shanghai, China. All used chemicals were purchased from a local chemical store at Shanghai and used as received.

The electro-synthesis process was performed as Figure 7-47 described by employing an electrostatic generator to assist the common solution polymerization by taking the positive and negative electrodes alternatively to link either the ANI solution (minus) or the outer wall of a stainless steel container (plus) to form two different electric cycles corresponding to the EI reduce or enhance, respectively. 0.45 mL of ANI and 0.32 g of citric acid were dissolved in 24 ml distilled water to form an ANI/acid/water solution at

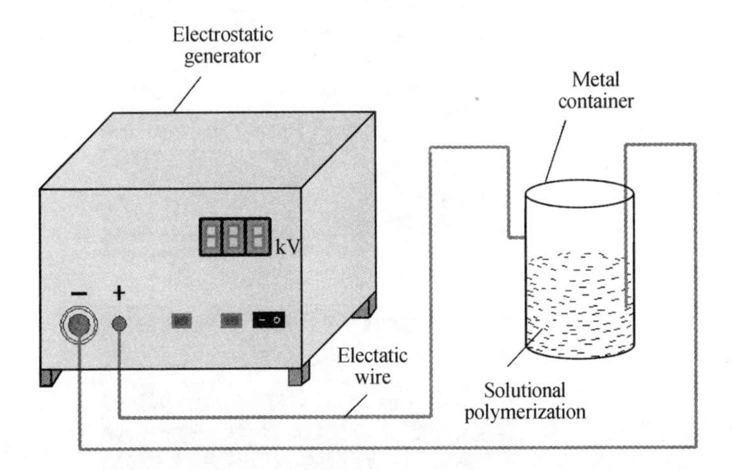

Figure 7-47 Scheme on the electro-synthesis of PANI nanofibers. In this experiment, the positive and negative electrodes were alternatively linked to either the ANI solution or the metal container to form two electric cycles corresponding to electrostatic interactions enhance or reduce, respectively

a constant pH of 4.0. The aqueous solution of APS prepared by dissolution of 1.09 g APS in 12 mL of distilled water was then rapidly added to the ANI solution under vigorous stirring condition for about 1 min, 25 ℃. After synthesis, the dark green PANI powder was washed by deionized water and ethanol, alternatively, for several times until the filtrate presented colorless. The sample was finally vacuum oven dried at 60 ℃ for 24 h. An electric-free sample was prepared as a reference using the normal solution polymerization.

FESEM images of two electro-synthesized PANI nanofibers and referenced sample were showed in Figure 7-48. A comparison of these three samples interesting found that the electric-free sample (Figure 7-48, top) showed only aggregates where no fiber shape while the EI reduce (Figure 7-48, middle) or enhance (Figure 7-48, bottom) both showed nanofiber shape and size. In terms of Figure 7-48, two electro-synthesized PANI nanofibers both have similar average diameter, d_a, at about 100 nm and their difference is in the length because the EI reduce formed shorter length in the range of 200～300 nm (Figure 7-48, middle), and the EI enhance formed longer length greater than 1 μm (Figure 7-48, bottom). This suggested that the electro-synthesis method is not only capably for fabrication of PANI nanostructure, but also provided two possibilities for controlling the nanosize by alternating the electrode linkage.

Synthesis condition	SEM images of PANI	Comment
Normal solution synthesis		No fiber shape
An EI reduce-based electro-synthesis by taking the positive and negative electrodes linking to the ANI solution and metal container, respectively	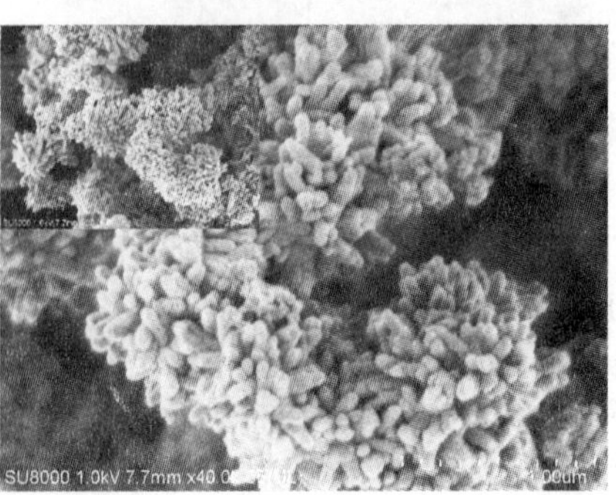	d_a = 100 nm L = 200～300 nm

Synthesis condition	SEM images of PANI	Comment
An EI enhance-based electro-synthesis by taking the positive and negative electrodes linking to the metal container and ANI solution, respectively	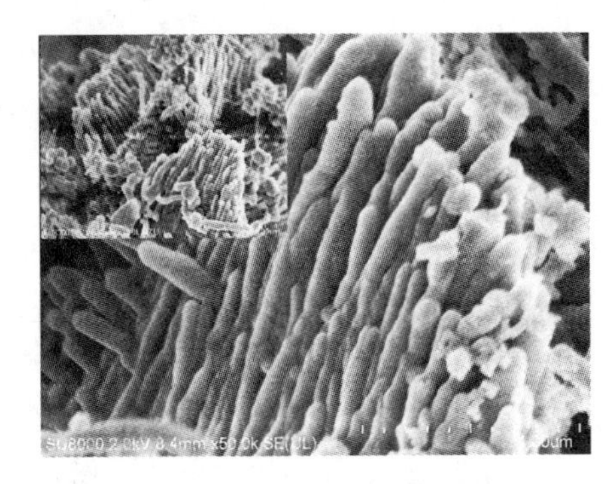	$d_a = 100$ nm $L \geqslant 1$ μm

Figure 7-48 **FESEM images of normal PANI sample (top) and electro-synthesized PANI nanofibers in relation to the electrostatic interactions, EI, reduce (middle) or enhance (bottom), respectively**

In fact, Figure 7-48 showed that the electro-synthesized PANI nanofibers appeared to be an array. This regular 2D form is interested because the application of conducting polymer required such forms to fit some cases as known from literature.

Figure 7-49 presented the FTIR spectra of three PANI samples. It was found all these samples showed intense peaks located at 3 436 cm^{-1} due to the N—H stretching, at 2 523 cm^{-1} due to the CH_3 stretching, at 1 804 cm^{-1} due to the C=O stretching, at 1 580/1 587 cm^{-1} due to the C=C stretching of the quinoid ring, at 1 503 cm^{-1} and 1 446/1 452 cm^{-1} due to the C=C stretching of benzenoid rings, at 1 297 cm^{-1} due to the C—N stretching of benzenoid unit, at 1 156 cm^{-1} due to the C—H aromatic in plan bending, at 1 048 cm^{-1} due to the quinoid ring, and at 876 cm^{-1} due to the 1,4-disubstituted aromatics. However, a comparison of these samples found that the normal PANI showed a unique shoulder peak at 1 477/1 568 cm^{-1} and two electro-synthesized samples both created a new peak to insert in this shoulder to appear as 1 446/1 503/1 581 cm^{-1}. This importantly implied that the PANI nanostructure formed by electro-synthesis is due to the enhancement of the C=C stretching of the quinoid ring.

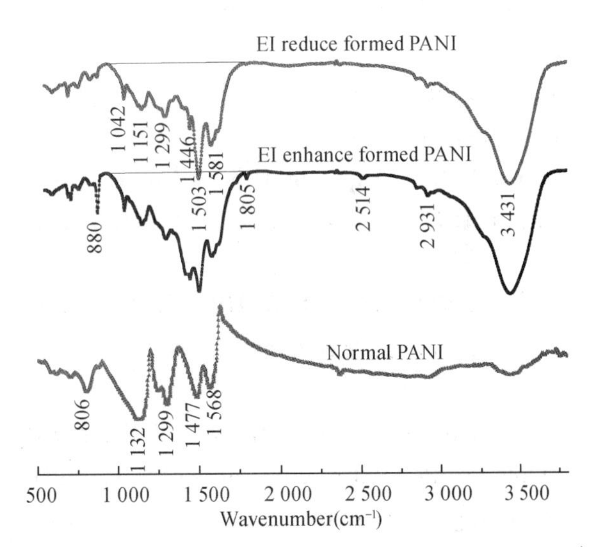

Figure 7-49 **FTIR spectra of normal and electro-synthesized PANI nanofibers in relation to the electrostatic interactions enhance or reduce, respectively**

Additionally, it is furthermore found that the shoulder peak at 1 446 cm^{-1} has been enhanced about 57% for EI enhance-based PANI as compared with the EI reduce-formed sample. This quantitative difference is reasonable to suggest that the EI enhance (Figure 7-48) formed PANI would be strongly in the C=C stretching deformations of the benzenoid rings.

The XRD patterns of three PANI samples were showed and compared in Figure 7-50. Observe the common PANI showed intense 2θ peaks at 20.6° and 25.4° corresponding to the reflection planes of (020) and (200) periodicity parallel and perpendicular to the polymer chains, respectively, while two electro-synthesized PANIs both showed two intense shoulder peaks at 18.7°/19.7° and 25.8°/26.6°, respectively. Obviously, the former shoulder peak was derived from the 20.6° and the latter was derived from the 25.4°, respectively. Since two electro-synthesized PANI samples both showed a new peak at 23.4° (Figure 7-50), this suggests that the amorphous area was increased for electro-synthesized sample as compared with the normal PANI.

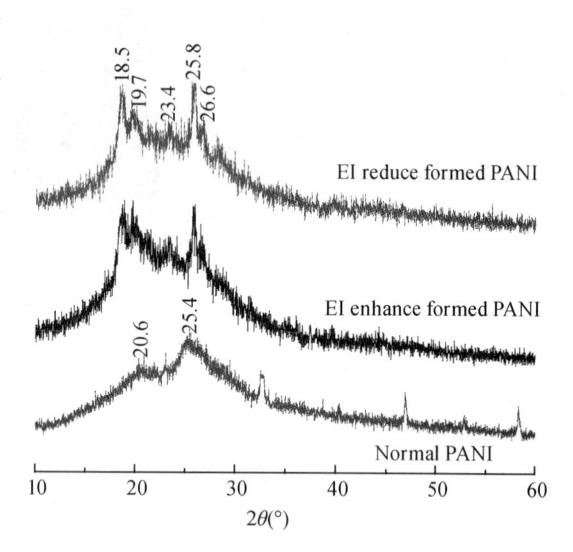

Figure 7-50 XRD patterns of normal and electro-synthesized PANI nanofibers in relation to the electrostatic interactions enhance or reduce, respectively

7.8 Case on directly formation of PANI nanotube by various polymerizations

Polyaniline, PANI, is one of the mostly studied intrinsically conducting polymers owing to its better environmental stability, tunable conductivity switching between insulating and semiconducting materials, and reversible redox behavior. PANI nanostructureis greatly expected due to its nano-effect fitting some novel applications and of which the nanotube is an interest. Cyclodextrin, CD, is a family of compounds made up of sugar molecules bound together in a ring and cyclic oligosaccharides. Since typical CD contains a number of glucose monomers ranging from six to eight units in a ring, creating a cone shape and denoted as the α-CD, β-CD and γ-CD, respectively (Figure 7-51). Recently, we have studied the adsorption properties of the α-, β- and γ-CDs and found they can absorb the non-polar liquids greatly as compared with the polar liquids, and their Lewis acid-base interaction properties are varied to follow an order such as: γ-CD > β-CD > α-CD. Therefore, to take this aspect of three CDs as three templates to prepare PANI nanotubes is proposed.

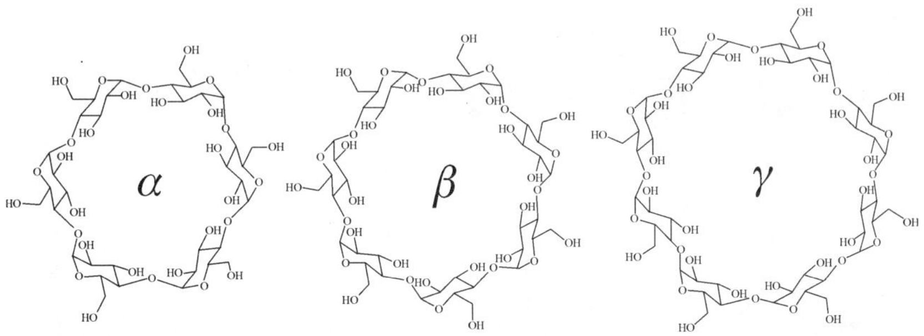

Figure 7-51 A description of the structure of α-CD, β-CD and γ-CD, respectively

7.8.1 Directly formation of PANI nanotubes with varied pore shape by solution polymerization

The aniline (99%) and ammonium peroxodisulfate, APS (99%) both obtained from the Sinopharm Chemical Reagent Co., Ltd. located at Shanghai, China. The commercial α-, β- and γ-CDs are obtained from Majorbio Biotech Co. Ltd. Shanghai, China and used as received.

Synthesis of PANI nanotubes was performed via a solution polymerization. A typical synthetic process was employed as follows: 0.45 ml of aniline (ANI) and 0.32 g of citric acid were initially mixed in 24 ml distilled water to form the mixture solution at a constant pH of about 4.0. The aqueous solution of APS (1.09 g of APS in 12 mL of distilled water) was then rapidly added to AN/acid solution under vigorous stirring condition for about 1 min. Following the stirring, 1 mmol CD was added into solution and the resulting mixed solution was left for 24 h and kept at 4℃ in an ice box. Finally, the obtained PANI powder was obtained and washed with deionized water for several times until the filtrate presented colorless then vacuum oven dried at 60℃ for 24 h.

FESEM images of three PANI nanotubes corresponding to three CDs are showed in Figure 7-52 left. Observe that all these samples showed nanotube structure, however, in different pore shapes. For example, the pore shape for PANI/α-CD is in flat (Figure 7-52, top), for PANI/β-CD is in rectangle (Figure 7-52, middle) and for PANI/γ-CD in triangle (Figure 7-52, bottom), respectively. Obviously, these differences are caused by those used CDs due to their varied glucosemonomers. In Figure 7-52 right, the pore diameter determined using BET method is also showed and the greatest is found for PANI/β-CD, and the smallest is for PANI/α-CD. This order is interested because it without following the order of the glucosemonomers of these CDs. In order to understand the reason why the pore size of these PANI nanotubes without following the order of CDs.

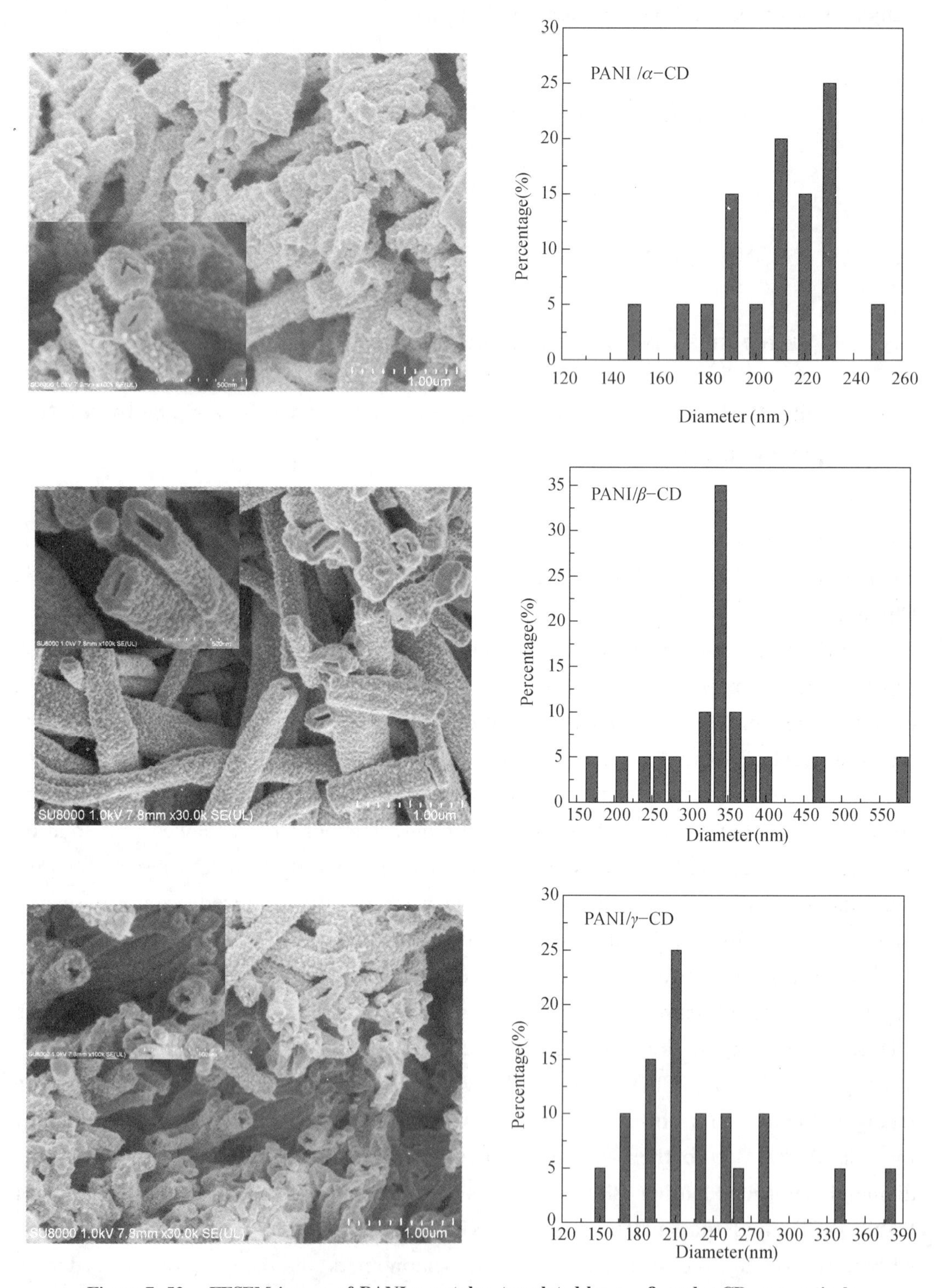

Figure 7-52 FESEM images of PANI nanotubes templated by α-, β- and γ-CD, respectively

In terms of their glucose monomers, the average pore diameter, d_a, was taken as a function of the surface property parameters, e. g. the Lifshitz-van der Waals interaction component, γ^{LW} (a plus of the dispersion force of London, the orientation force of Kessom and the dipole-induced dipole force of Debye), the Lewis acid component, γ^{+}, of CDs, the specific surface area, S, and the conductivity of those three CDs as four plots showned in Figure 7-53, respectively. Noted that the d_a of these nanotubes is increased with both the γ^{LW} and γ^{+} of these CDs (Figure 7-53 top) to indicate that these PANI nantubes are self-assembled in accordance with literature and driven by these forces, especially the γ^{LW}. In fact, this also indicated that the pore shape of these PANI/CD nanotubes is dominated by the γ^{LW} of CDs. According to Figure 7-53 (top), the d_a of PANI nanotubes is also slightly influenced by the γ^{+} of CDs due to the doping of Cd-induced an increase of the Lewis acid for PANI in agreement with literature. In Figure 7-53, the increase of the averaged diameter, d_a, is found to increase the specific surface area, S, and reduce the conductivity of these three PANI nanotubes. These are reasonable because a great d_a should have a big S, while the conductivity should be on the contrary.

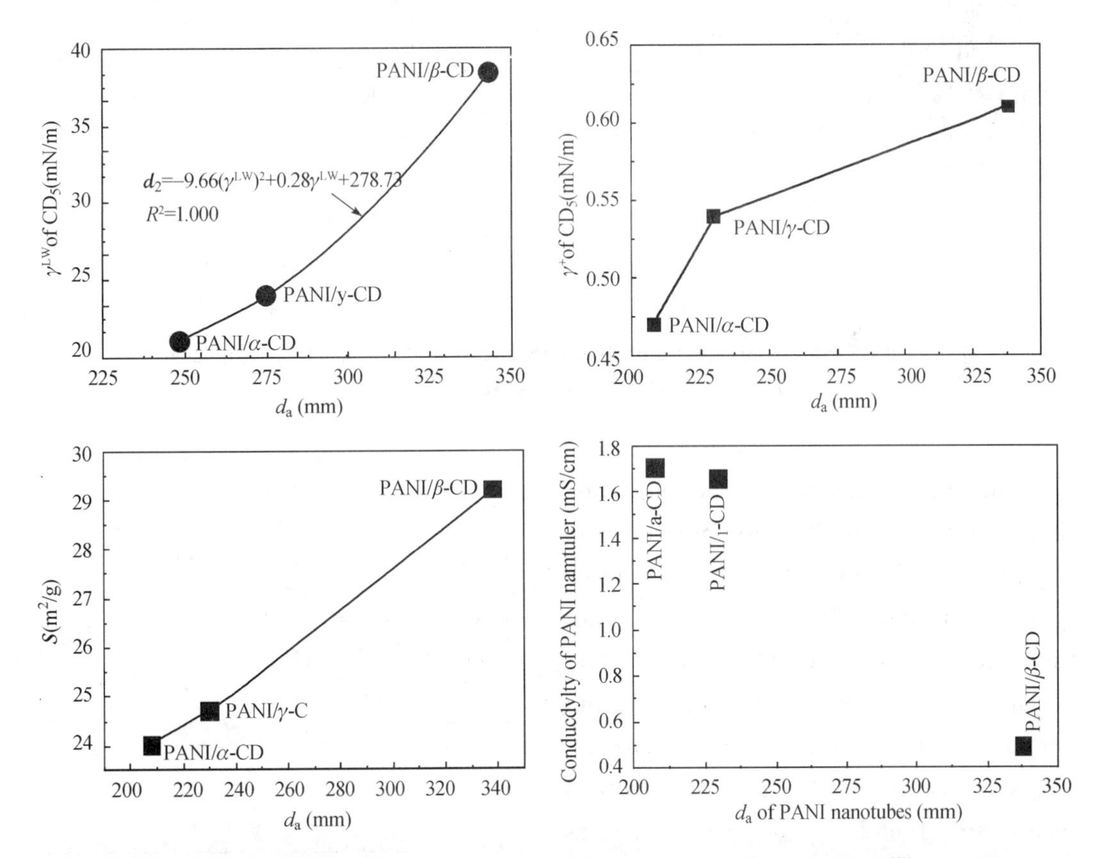

Figure 7-53 **The average pore diameter, d_a, as a function of the LW component, γ^{LW} (top left), the Lewis acid component, γ^{+}, the specific surface area, S (bottom left), and the conductivity of PANI nanotubes template by α-, β- and γ-CD, respectively**

FTIR spectra of pure and three PANI nanotubes were compared in Figure 7-54(A), where the typical PANI peaks appeared at 1 140 cm^{-1} and 1 301 cm^{-1} due to the aromatic

C—H in-plane bending and the C—N stretching of secondary aromatic amine, at 1 495～1 575 cm^{-1} associated with benzenoid rings and the C=C stretching deformation of quinoid. All these PANIs also presented main peaks at 3 432 cm^{-1} corresponding to the N—H stretching mode, at 2 920 cm^{-1} due to the C—H stretching. The main CD peaks appeared at 3 315～3 337 cm^{-1}, 1 642～1 658 cm^{-1}, 1 153 cm^{-1}, 1 024～1 028 cm^{-1} (Figure 7-54A) and are overlapped in PANI/CDs (Figure 7-54, B). This suggests the PANI and CD formed the host-guest inclusion complex. In fact, the CDs contributed two new peaks in the spectra of three PANI/CDs at 997 and 970 cm^{-1} (Figure 7-54, A) indicating its template role.

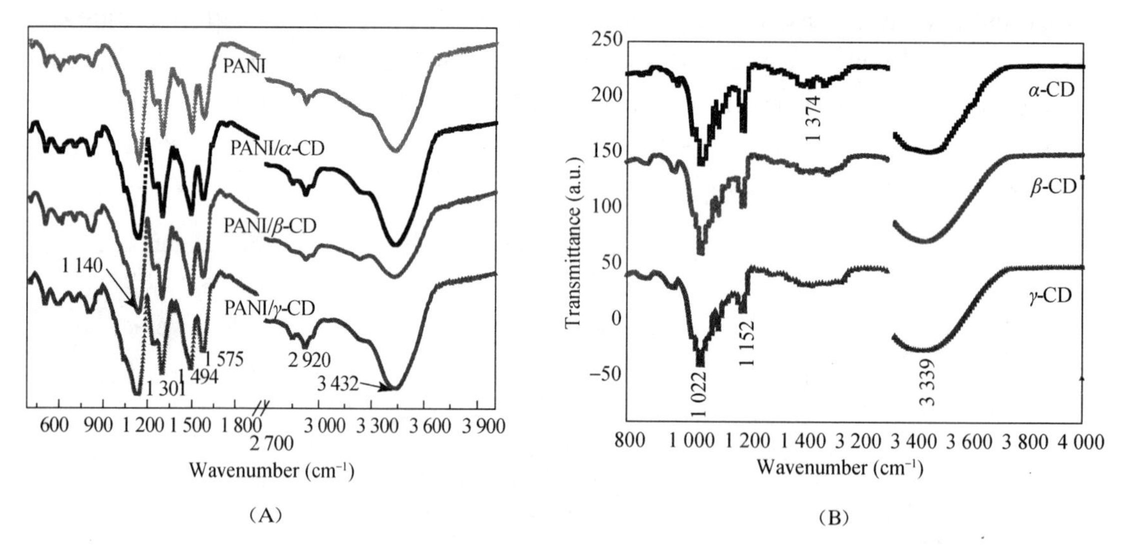

Figure 7-54 FTIR spectra of three CDs and three PANI nanotubes templated by α-, β- and γ-CD, respectively

The XRD patterns of the pure and three Cd-based PANIs were presented in Figure 7-55. The pure PANI appeared two typical 2θ peaks at 19.1°～25.5° represent the reflection planes of (020) and (200) because of the periodicity parallel and perpendicular to the polymer chains, respectively. In PANI/CDs, two typical peaks are found to be passivated as compared to the sharp peaks at pure PANI to suggest the formation of the host-guest inclusion complex accompanied by reducing the crystallinity.

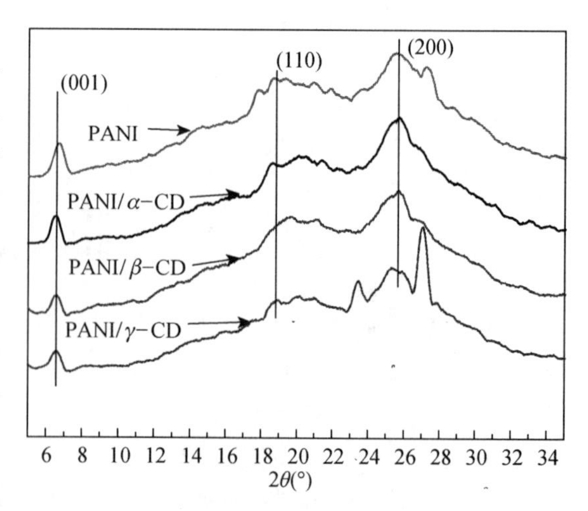

Figure 7-55 XRD patterns of pure PANI and three PANI nanotubes templated by α-, β- and γ-CD, respectively

7.8.2 Directly formation of PANI nanotubes with controlled square pore structure by solution polymerization

PANI nanostructure is greatly expected due to its nano-effect fitting some novel applications and has been highlighted and extensively studied in relation to the use of hard, soft or free-template in polymerization of aniline. Among various nanostructures, PANI nanotube has aroused wide attention because it is a plus of the carbon nanotubes. Formation of PANI nanotubes with rectangle pore shape was interested by researchers. *Wan* et al. have developed a template-free method. *Stejskal* et al. found that the nano oligomer crystallites served as the starting templates for the nucleation of PANI nanotubes. *Gao* et al. synthesized the rectangular PANI nanotubes using oligomers as template. Yoshida et al. have studied the PANI/CD complex and found the cavity of CD suitable to host for polymer chain to constituted an inclusion complex. Zhou et al. found that bilayer-micelles can be served as the template to form rectangular nanotubes. Zhu et al. applied camphorsulfonic acid as a dopant to fabricate PANI nanotubes with rectangle shape.

We have applied cyclodextrin, CD, with α, β and γ-configuration to form PANI naotubes and found the β-CD guided PANI nanotube presented rectangle pore shape. However, to control this rectangle or square pore shape for PANI nanotubes is unknown. In this work we varied the ratio of added ANI and β-CD, and obtained three relative samples as Table 7-20 described with various pore structures and parameters.

Table 7-20 Parameters of PANI/β-CD nanotubes in relation to various added CD amounts

PANI/β-CD	Total added mass (%)	ANI (mmol)	β-CD (mmol)	K (S/cm)	S (m^2/g)
A	100	5.00	1.00	4.88×10^{-4}	24.0232
B	75	3.75	0.75	2.07×10^{-4}	30.1325
C	50	2.50	0.50	8.97×10^{-5}	26.1521

The FESEM and TEM images of these three PANI/β-CD nanotubes were showed in Figure 7-56, respectively. To take their concentrations of ANI and β-CD, especially the total added mass (Table 7-20) as a consideration, it was found that these nanotubes have changed pore shape and size and the applied concentration is a key for controlling the pore shape change. According to Figure 7-56, to reduce the total added amount of ANI and CD from 100% to 50% would lead the nanotubes initially keeping the rectangle pore shape but to increase the pore size then changing to the square pore shape meanwhile also changing the wall thickness.

Figure 7-56　FESEM images of PANI nanotubes templated by β-CD with various added amounts, respectively

FTIR spectra of pure PANI and PANI/CD were compared in Figure 7-57. The main characteristic bands of PANI were found to appear at 1 575 cm^{-1} and 1 488 cm^{-1} due to the C=N and C=C stretching deformation of quinoid and benzenoid rings, respectively, at 1 293 cm^{-1} is the C—N stretching mode for benzenoid ring and at 1 134 cm^{-1} is the C—H plane deformation in the 1, 4-dissubstituted benzene ring. Noted the bands of PANI/β-CD at 1 575 cm^{-1} and 1 293 cm^{-1} are slightly shifted to lower wavenumber as compared with the pure PANI to indicate the hydrogen bonding formed between the PANI chains and CD to dominate the nanostructure formation.

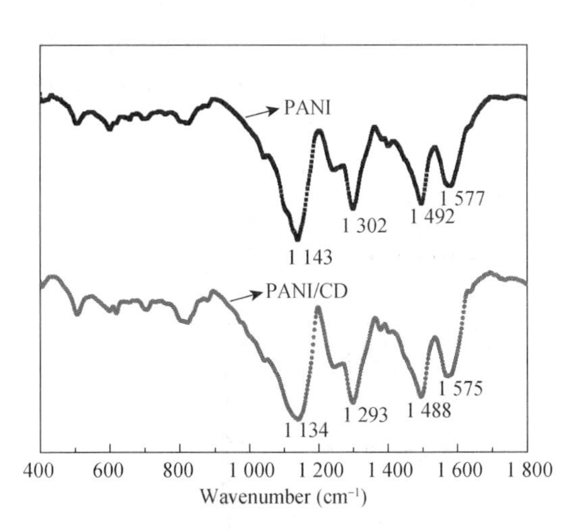

Figure 7-57 FTIR spectra of pure PANI and PANI/β-CD nanotube

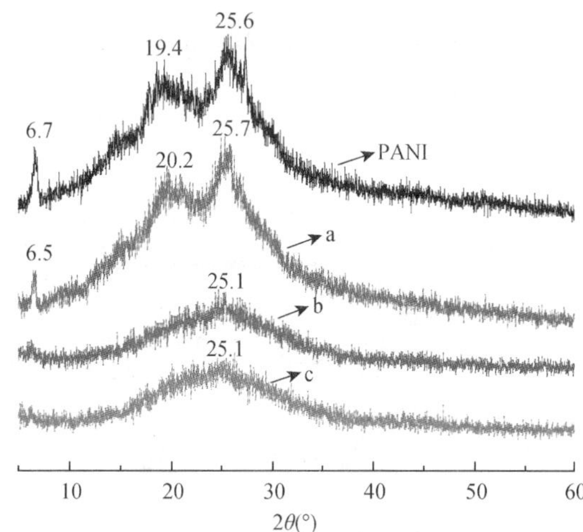

Figure 7-58 XRD patterns of pure PANI and PANI/β-CD nanotube

The wide-angle XRD patterns of pure PANI and three PANI/CD nanotubes were presented and compared in Figure 7-58. The sharp peak appeared at about 6.5° was found for pure PANI assigned to the periodicity distance between the dopant and the N atom of PANI on its adjacent main chains. Since this peak was disappeared for two PANI/CD nanotubes as can be seen in Figure 7-58, this suggests that the orientation of PANI chains along the tubular direction of PANI/CD would be occurred at low total concentration, e.g. less 75% (Table 7-20). This implied that the reduce of the total added masses to lower than 75% (Table 7-20) would cause PANI chains growth along the tubular direction as observed peaks at approximation of 19.4° and 25.6°, respectively, to cause the PANI chains periodicity parallel and perpendicular arranged. Since these peaks weakly appeared in Figure 7-58 than that of the others, this indicated that the oligomers might be more orderly arranged than that of the PANI macromolecules, especially in low total added masses because such condition can induce the PANI/β-CD inclusion complex.

The conductivity of these three PANI nanotubes is also presented in Table 7-20. To take these conductivity values to link Table 7-20 appeared morphology of those PANI nanotubes, it is clearly that the bigger rectangle pore has greater conductivity and the square pore has the smallest conductivity.

According to above analysis, a possible mechanism on the formation of rectangular or square PANI nanotubes with controlled pore shape and size is described as showed in Figure 7-59. Because of the hydrophobic internal cavity, the aniline doped with citric acid would incline to load on the internal surface of β-CD. Thus, owing to the interaction between PANI oligomers and β-CD, certain lamellar micelles would be formed by the host-guest inclusion complex at the early stage, and this would lead the secondary growth of the oligomers to be acted as the seeds to form the flat oligomer flakes, and finally the flakes united to form the tube structure due to the reaction of the active center on the edges causing the decreased concentration of reactants enlarged the distance between interactive groups longer.

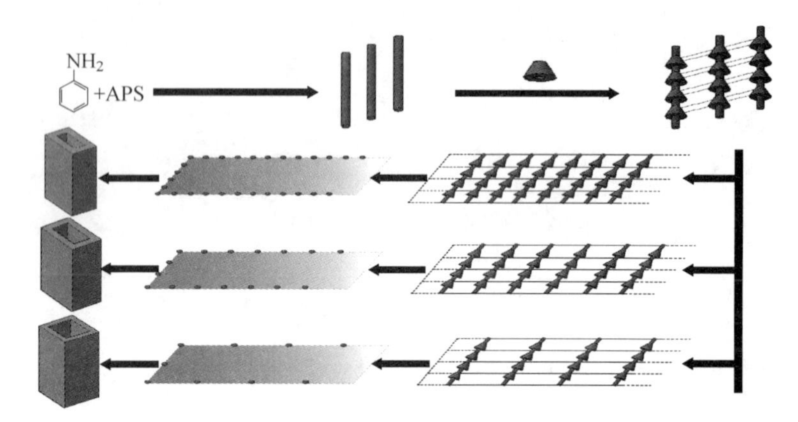

Figure 7-59　Mechanism on formation of pore shape and size controllable PANI nanotubes using β-CD as template

7.9 Case on directly formation of PANI micro/nano fiber by taking PLA as guider via solution polymerization

Since the finding of PANI, a lot of methods have been developed to synthesis of PANI micro/nanostructures with controlled morphology, e. g. as nanowires, rods and tubes, in relation to the aid of either the hard and soft template, e.g. the former to be the zeolite channels, track-etched polycarbonate and anodized alumina, and the latter to be the surfactants, micelles, liquid crystals, thiolated cyclodextrins and polyacids. It is also known that the possible methods are the electrochemical methods, electrospnning, mechanical stretching, interfacial polymerization, oligomer-assisted seeding method, emulsion, solution polymerization and electro-synthesis.

In this work we demonstrated that the use of chiral poly(L-lactic acid), PLLA, and poly(d-lactic acid), PDLA, can control the PANI structure. Since PLLA and PDLA have opposite chirality, their opposite effects are of interest.

The aniline (99%), ammonium peroxodisulfate, APS (99%) and citric acid were obtained from the Sinopharm Chemical Reagent Co., Ltd. located at Shanghai, China. All used chemicals were purchased from a local chemical store at Shanghai and used as received.

PLLA (L149) and PDLA (D155) both with the same molecular weight at about 11.6×10^4 supplied by Zhejiang Hisun Biomaterials Co. Ltd. were used as received.

PLLA PDLA

Figure 7-60 The structure of PLLA and PDLA

The interfacial synthesis of PANI/PLA samples was performed by preparation of two phases as Figure 7-61 described where the top was an aqueous phase, AP, formed by dissolution of 1.09 g APS in a solution of 0.32 citric acid/100 mL water, and the bottom was an oil phase, OP, formed by dissolution of 0.27 g PLA in a mixture of 0.45 mL ANI/100 mL $CHCl_3$.

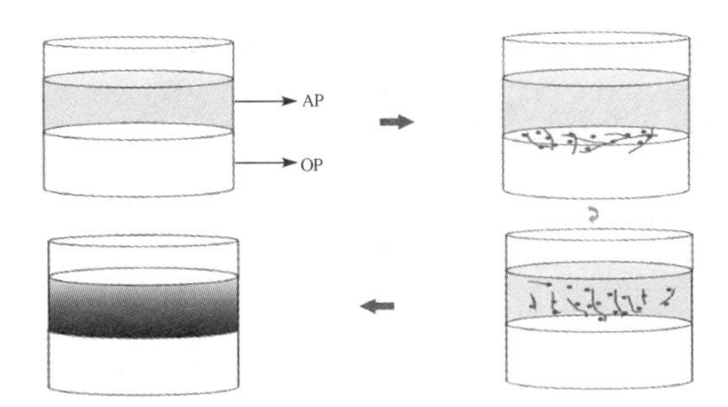

Figure 7-61 Scheme on interfacial polymerization of PANI in the presence of PLA, where the PLA and ANI dissolved in organic solvent to form the oil phase, OP, and the APS solution as the aquaous phase, AP

Each polymerization run was performed at 0 ℃. After polymerization, the PANI powder was washed with deionized water and ethanol for several times until the filtrate presented colorless then be vacuum oven dried at 60 ℃ for 24 h.

The effect of the chirality of PLLA and PDLA on the morphology of PANI was observed in Figure 7-62. It is interesting to find that the morphology of these PANIs doped by PLLA showed a vertebra-like 1d-microstructure with an averaged diameter at about 500 ~ 700 nm and by PDLA showed a joint-like 1d-microstructure with a n averaged diameter at about 400~500 nm.

PLLA-controlled PANI (d_a = 500 ~ 700 nm), ANI/PLLA=0.45 mL/0.27 g	PDLA-controlled PANI (d_a = 400 ~ 500 nm) ANI/PDLA=0.45 mL/0.27 g
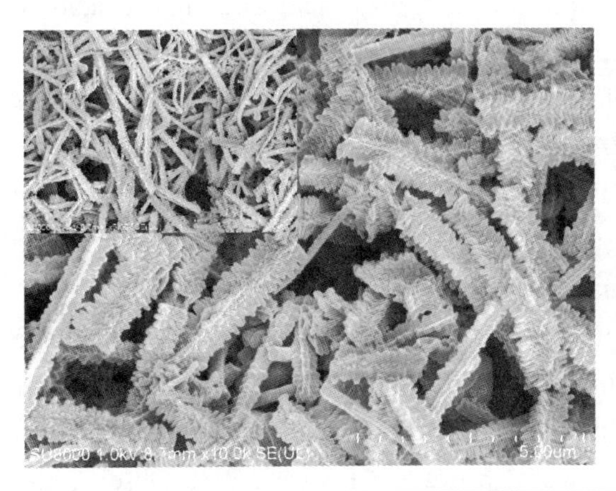	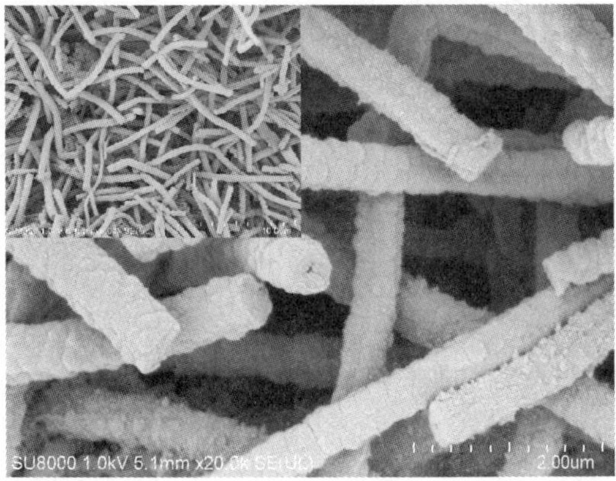

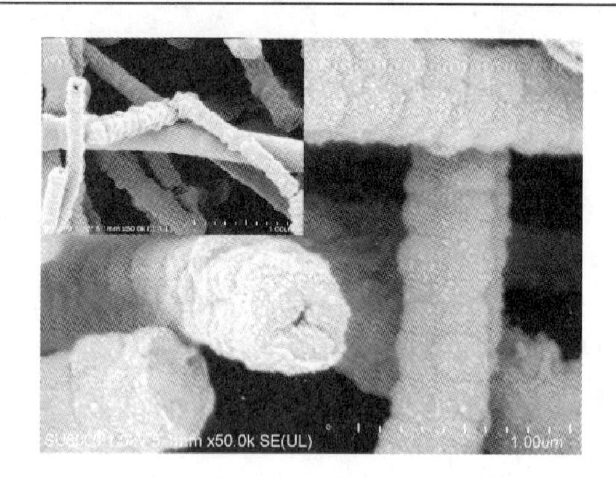

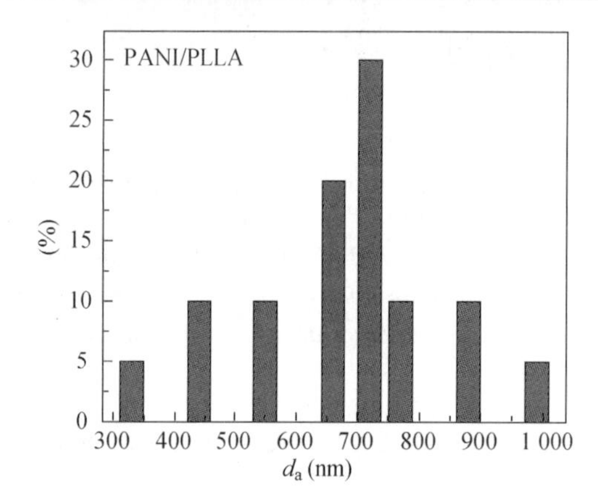

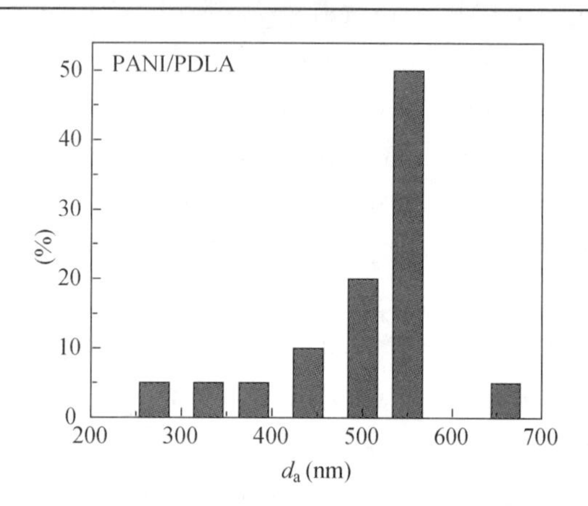

Figure 7-62　FESEM images of PANI templated/doped by PLLA (left) or PDLA (right), respectively

These differences are interested and obviously correspond to the role of these two PLA-based chiral aspects, e.g. the L- and d-type. Since the PANI morphology doped by PLLA seems to like those doped by acetic acid, hexanoic acid or molybdic acid, while doped by PDLA seems to be similarly as those doped by lauric acid or stearic acid. These suggested that the PLLA may have the similar role as acetic acid, hexanoic acid and molybdic acid and PDLA may have the similar role of lauric acid and stearic acid to control the morphology of PANI. This finding is interested because PLA is a bio- and degradable polymer unlike those low molecular weight acids. In other words, this indicated that we can apply PLA to dope PANI to form new conducting materials.

On these PLLA and PDLA controlled PANIs, it is probably that they may contain high density surface/interface to induce deep energy levels within the band gap and trap the free or photoexcited carriers from the inside.

In order to understand the doped amount of PLA how to influence and control the morphology of PANI, a series of PANI samples were prepared by varying the ratio of ANI/PLLA (mL/g). The related FESEM images of the morphology of those PANIs in relation to used various ratios of ANI/PLLA were showed in Figure 7-63. Observe the cross-section of those PANI samples all showed four leaves morphology while the shape was evidently changed with the ANI/PLLA ratio changes due to the increase of the PLLA amount because the ratio of PANI/PLLA (mL/g) at 0.450/0.135 appeared four round leaves (Figure 7-63, top) and at 0.450/0.270 changed to non-round form (Figure 7-63, middle) and at 0.450/0.540 changed to four fork-like leaves (Figure 7-63, bottom).

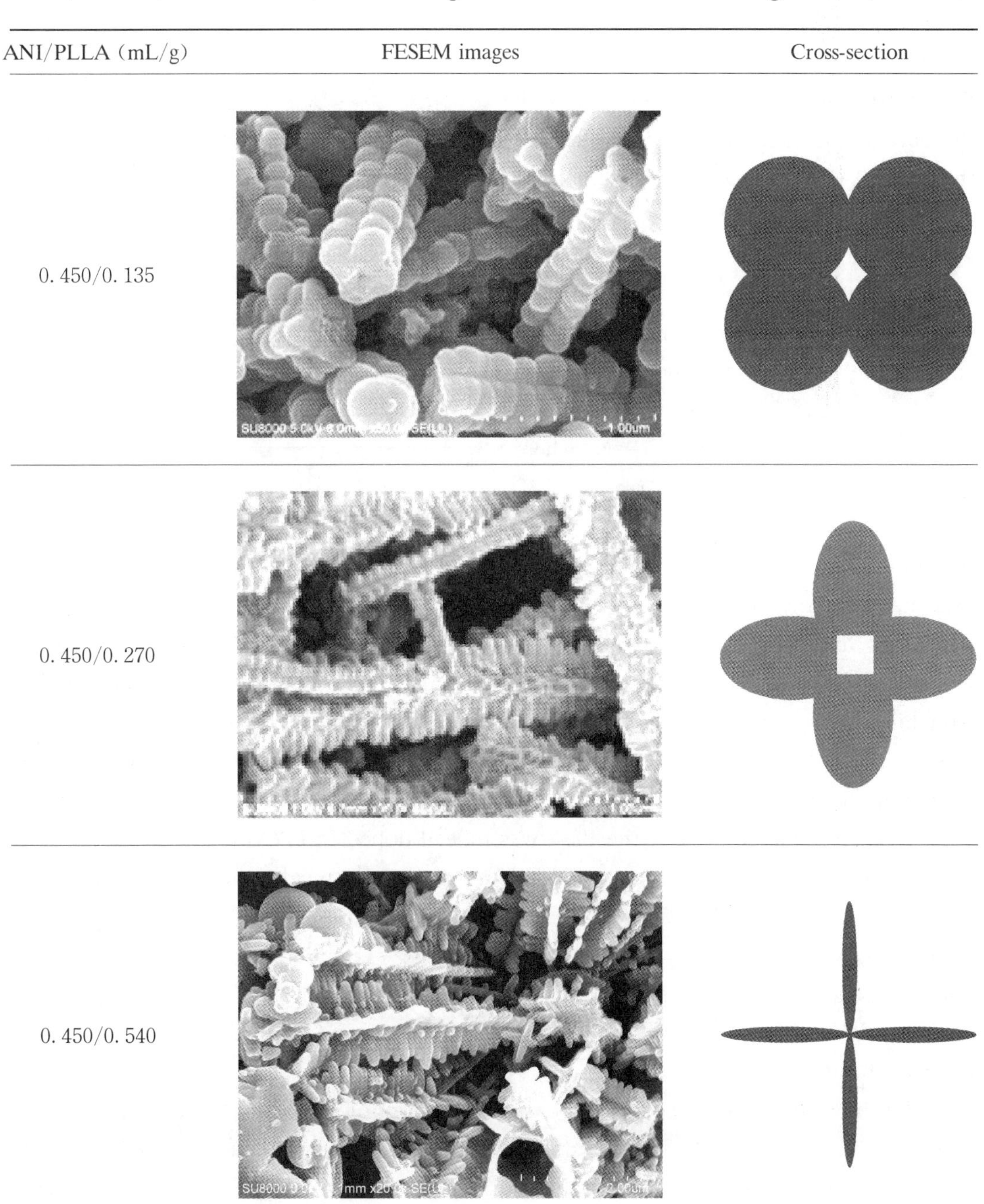

Figure 7-63 FESEM images of the morphology of PANI templated/doped by PLLA in relation to various ANI/PLLA ratios

Obviously, this series changes indicated that the PLA is indeed a dominator of PANI morphology and the variety of doping amount can precisely control the PANI morphology. In fact, Zhang et al. have also found similar phenomena that the varying of the ANI/acid ratio can adjust the morphology of PANI. Since the increase of the PLLA caused the leaves in sharp (Figure 7-63, bottom), it is considered that one can form rough or smooth PANI surface to fit different application requests.

FTIR spectra of two PLA-doped PANIs were showed and compared with the pure PANI in Figure 7-64. Observe the pure PANI showed intense bands at 3 440 cm^{-1} due to N—H stretching vibration, 3 240 cm^{-1} and 2 920 cm^{-1} due to CH_3 stretching, 1 586 cm^{-1} due to C=C stretching of the quinoid rings, 1 504 cm^{-1} due to C=C stretching of benzenoid rings, 1 310 cm^{-1} due to C—H stretching mode, 1 157 cm^{-1} due to N=Q=N, where Q represents the quionoid ring, 1 042 cm^{-1} and 827 cm^{-1} due to 1, 2, 4-substituted aromatic ring due to the existence of the cross-linked network structures, and two PLA-doped PANIs both consistently showed some intense bands without overlap to the pure PANI, e. g. at 3 006 cm^{-1} and 2 964 cm^{-1} due to CH_3 stretching, 1 777 cm^{-1} and 1 749 cm^{-1} due to C=O stretching, 1 468 cm^{-1}, 1 443 cm^{-1}, 1 396 cm^{-1} and 1 381 cm^{-1} due to CH_3 stretching, 1 222 cm^{-1} due to COC and CH_3 stretching, 1 144 cm^{-1} due to CH_3 stretching and 1 053 cm^{-1} due to C—CH_3 stretching. This reasonably indicated that the PLA was indeed doped in PANI chains.

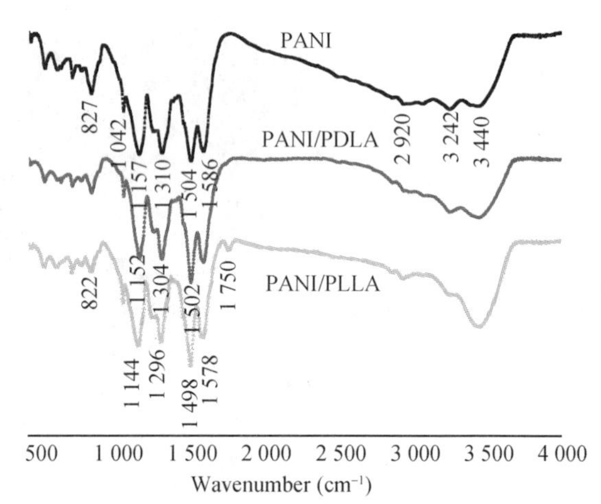

Figure 7-64 FTIR spectra of PANI templated by PLLA and PDLA, respectively

A comparison of these three samples found that the typical shoulder peaks at 1 586/1 504 cm^{-1} in pure PANI was shifted to about 1 578/1 498 cm^{-1} for two PLA-doped PANIs, especially in narrow for PLLA-doped PANI (Figure 7-64). This difference is interested because it suggests that the quinoid and benzenoid rings of PANI due to the electrons delocalization were strongly influenced by doping of PLLA due to its crystal polymorphism.

This therefore explained that the formed morphology of PLLA-doped PANI as Figure 7-63 showed is due to the doping of L-type chiral structure caused the CH_3 stretching enhancement to lead the formed PANI having the induced crystal polymorphism. Additionally, since the PLLA-doped PANI showed a unique band at 1 750 cm^{-1} corresponding to the C=O stretching, this furthermore indicated that the PLLA molecules are mainly doped in PANI chains than that of the PDLA. This thus further explained why the PLLA-doped PANI showed vertebra-like form with four leaves while the PDLA-doped PANI appeared only a joint-like form.

The XRD patterns of two PLA-doped PANIs were showed in Figure 7-65, where the pure PANI was also appeared as a reference. Observe the pure PANI showed typical 2θ peaks at 20.9° and 24.7° representing the reflection plane of (020) and (200) corresponding to the periodicity parallel and perpendicular to the polymer chains, respectively. However, the PANI doped by PLLA showed peak shifting approaching a narrow tendency because the former peak was right shifted to at 21.5° and the latter peak was left shifted to at 24.1°, respectively, and the PANI doped by PDLA showed only the latter peak shift, e. g. from 24.7° to 23.6°. These differences again indicated that the PLLA was mainly doped in PANI chains than that of PDLA doping and the doping of PLA in PANI chains caused the stereocomplex crystallization in a trigonal unit cell of dimensions.

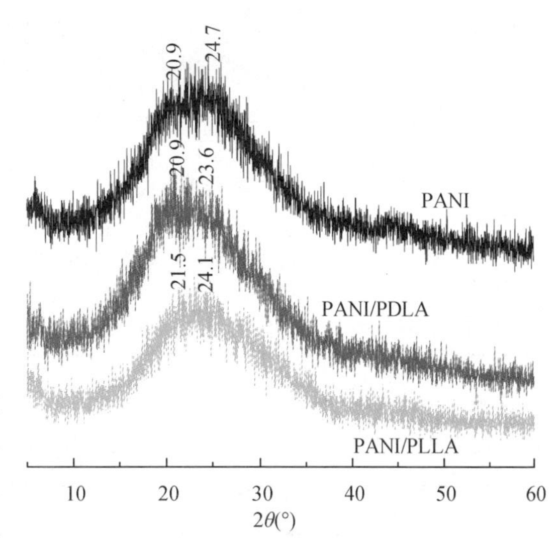

Figure 7-65 XRD patterns of PANI template by PLLA and PDLA, respectively

7.10 Case on directly formation of PANI nanofiber using LGS as dopant via solution polymerization

As previously described that the use of lignosulfonate, LGS, in polymerization of PANI the LGS played as a dopant. In fact, LGS can play the role as template to form PANI nanofibers.

The aniline (99%), ammonium peroxodisulfate, APS (99%) and citric acid were obtained from the Sinopharm Chemical Reagent Co., Ltd. located at Shanghai, China. A commercial LGS-Na from Jiangmen Sugar Cane Chemical Factory, Guangdong of China, was used as received. It has averaged diameter at (611 ± 50) nm. All used chemicals were purchased from a local chemical store at Shanghai and used as received. Lab made distilled water was always used through the whole work.

The synthesis process was performed as started by mixing 0.45 mL of aniline and 0.32 g of citric acid in 24 mL distilled water to form the mixture solution at a constant pH of about 4.0. The aqueous solution of APS (1.09 g of APS in 12 mL of distilled water) was then rapidly added to AN/acid solution under vigorous stirring condition for about 1 min. 0.1 g or 0.05 g LGS was then added into above solution to yield mixed solution which was left for 24 h and kept at room temperature, 25 ℃. Finally, the PANI powder was washed with deionized water and ethanol for several times until the filtrate presented colorless then be vacuum oven dried at 60 ℃ for 24 h.

Table 7-21 showed detailed concentrations on two prepared PANI/LGS nanofibers.

Table 7-21 Polymerization condition and parameters of PANI nanofibers templated by various LGS

PANI/LGS	ANI (g)	LGS (g)	Conductivity (S/cm)	DMF (mg/L)	DMSO (mg/L)	NMP (mg/L)
PANI	0.50	0	1.28×10^{-4}	0	0	0
PANI/LGS-2	0.45	0.05	4.87×10^{-3}	12.4	8.1	3.0
PANI/LGS-1	0.40	0.10	3.38×10^{-3}	/	/	/

The FESEM images of PANI samples without or templated by LGS were showed in Figure 7-66. Of which, the added LGS amount was varied as Table 7-21 described in

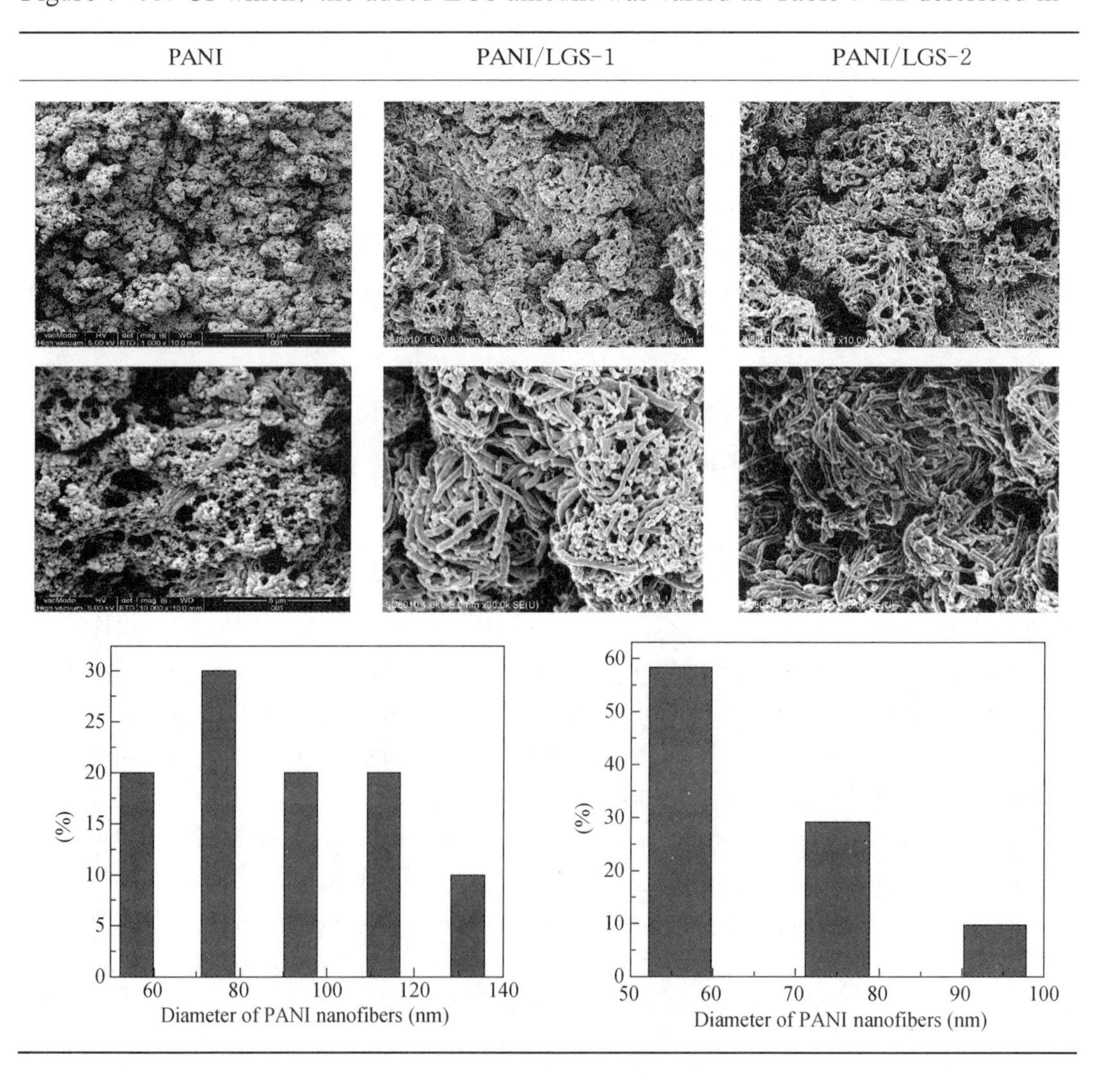

Figure 7-66 FESEM images of PANI/LGS nanofibers

detail. A comparison of three samples found that the pure PANI showed non-fiber type while two LGS-doped PANI both in nanofiber shape. In fact, according to fiber size analysis as seen in Figure 7-66 bottom, two LGS templated PANI nanofibers has average diameter less 100 nm.

The LGS is indeed a dopant in this case because two LGS-based PANIs both showed enhanced conductivity as compared to the pure PANI and enhanced the solubility of PANI (Table 7-21).

7.11 Case on formation of lignin-based carbon fiber

7.11.1 Formation of lignin-based carbon nanofiber

Carbon nanofiber, CNF, is an important material due to its chemical, electrical, magnetic and mechanical properties. CNF can be produced from inorganic or organic materials, and from the latter is mostly the polyacrylonitrile, PAN, and lignin. According to literature, the diameter of PAN-based CNF can be ranged within 130～280 nm while it would be great for lignin-based CNF, e. g. 400 ～ 2 000 nm due to the inherent limitations. However, it is truly that the lignin is quite inexpensive and renewable as compared with the PAN. Therefore, to fabricate lignin-based CNF is required.

Lignin is usually obtained from pulping and which obtained from the alkaline pulping is defined as the alkali (kraft) lignin and from the organsolving pulping is defined as the alcell lignin. The latter is easily electrospun due to its purity higher than that of the alkali lignin. However, alkali lignin is the main industrial products because most pulping factories applying the alkali process. Previously, we have applied alkali lignin as raw materials prepared carbon film and activated carbon fiber, respectively.

The aim of this work is to fabricate alkali lignin-based CNF with small size via the traditional methods, e. g. melt spinning and thermal treatments. This is because the traditional methods can fit large scale production as compared to the electrospinning. Experimentally, we applied poly(butylene terephathalate), PBT, to blend with alkali lignin to form a blend fiber by melt extrusion and spinning, then the as-spun blending fiber was oxided and carbonized to convert into CNF. The obtained CNFs were characterized by scanning electron microscopy, differential scanning calorimetry, infrared and element analysis.

A commercial alkaline lignin obtained from Tokyo Chemical Industry Co., Ltd. was used as received. According to the producer, this lignin has the molecular weight, M_w of about 28 000 and M_n of about 5 000. The used PBT is also a commercial product obtained from Yizheng Chemical Fiber Co. China. According to the company, this PBT has the intrinsic viscosity of (0. 82 ± 0. 02) dL/g, the melting point greater than 250 ℃ and the melting index about 7. 0 g/min.

Initially, the lignin powders and the PBT chips were vacuum oven preheated, respectively. The oven condition for the former is at 65 ℃, 8 h in order to remove the moisture, and for the latter is at 80 ℃, 8 h in order to cause the pre-crystallization. Then, these two polymers were blended at different ratios, e.g. 25%/75%, 50%/50% and 75%/25% (lignin/PBT, weight/weight), by thermal extrusion at 230 ℃ using a twin-screw micro-compounder (DACA, Germany). Since the 25%/75% (w/w) lignin/PBT blend presented better mechanical properties and spinnability as comparing with others, this blend fiber was furthermore spun via a twin-screw compounder (Haake RC90, Germany) at 230 ℃ to form the as-spun blend fiber. This as-spun blend fiber was finally thermally treated by an air oxidation stage and a carbonization stage under nitrogen atmosphere, respectively. Four different thermal conditions were employed and the detailed parameters were summarized in Table 7-22.

Table 7-22 Conditions for fabrication of alkali lignin-based CNFs

CNFs	Oxidation		Carbonization	
	Heating rate (℃/min)	Temp. (℃)	Heating rate (℃/min)	Temp. (℃)
CNF-1	2	250	3	800
CNF-2				1 000
CNF-3	3			800
CNF-4				1 000

Because the applied production steps are similar than that of *Kadla* et al. reported and the main difference is this case we used PBT to replace the PET, to compare the fiber shape and size of our samples with those reported by Kadla's is interested. The SEM photographs of four samples defined as CNF-1, CNF-2, CNF-3 and CNF-4 corresponding to Table 7-21 presented conditions were showed in Figure 7-67, respectively. It was found that these samples clearly showed two types, e.g. the CNF-1 and CNF-3 in carbon blocks, and CNF-2 and CNF-4 in expected CNF. Since CNF-4 presented visible fiber form with fine sizes, e.g. the diameter ranged between 100～300 nm, and the length ranged between 10～15 μm, which obviously either smaller or longer than that of literature reported values also using the alkali lignin. This indicated that the condition of CNF-4 used is capable for producing alkali lignin-based CNF with small diameter via traditional methods. This is of importance because the traditional method can provide scaled production of CNF as compared with the electrospinning.

A comparison of CNF-2 and CNF-4 further found that the carbonization stage is of importance for forming alkali lignin-based CNF with smaller diameter and longer length. The reason is considered due to the CNF formed by high temperature-based carbonization stage to be related to recrystallization.

CNF-1　　CNF-2

CNF-3　　CNF-4

Figure 7-67　SEM images of CNF-1, CNF-2, CNF-3 and CNF-4

The IR spectra of four CNFs were illustrated in Figure 7-68, where the as-spun blend fiber was also showed as a reference. Since the referenced blend fiber presented several intense peaks at about 2 960 cm^{-1}, 1 720 cm^{-1}, 1 580 cm^{-1}, 1 264 cm^{-1}, 943 cm^{-1} and 727 cm^{-1} all contributed by the different non-carbon components while these peaks are obviously disappeared in four CNFs, it implied that the non-carbon components were gasified by applied two thermal treatment stages and the CNF was mainly contributed by carbon. This is also supported by a visible peak located at about 1 662 cm^{-1} due to the C—C structure stretching for all CNFs (Figure 7-68) while without showed in the as-spun blend fiber.

Considering Figure 7-67 presented two types of CNFs, to compare their IR spectra is necessary for understanding the related reasons. Since the carbon blocks (CNF-1 and CNF-3) presented a visible peak at 1 450 cm^{-1} assigning to the CH_3 groups stretching while the CNFs (CNF-2 and CNF-4) without showed this peak (Figure 7-68), it is known that the CNF was converted from the carbon block by enhancing the carbonization temperature to furthermore gasify the residue non-carbon components.

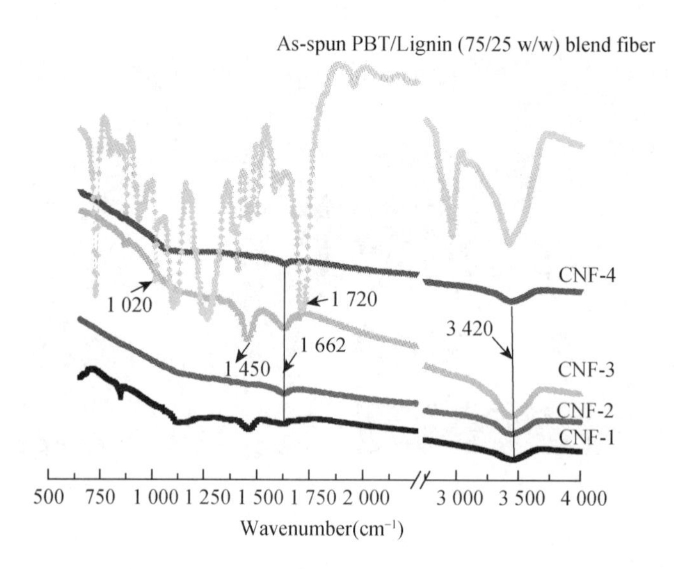

Figure 7-68　FTIR spectra of four CNFs and referenced as spun PBT/Lignin blending fiber

Considering only CNF-4 presented better fiber shape and size as compared with other prepared samples (Figure 7-67), its element analysis was performed and the results are showed in Table 7-23. Observed that this CNF-4 showed a high carbon percent, e.g. up to about 92% (Table 7-23) to support above findings.

Table 7-23　Elemental analyses of CNF-4.

Sample	C (%)	H (%)	N (%)	O (%)
CNF-4	91.88	0.82	0.29	7.01

The thermal properties of these four CNFs were analyzed by DSC as presented in Figure 7-69. Interestingly, these DSC curves also showed two types which not only corresponding to Figure 7-67 but also provided new insights because the two types of four CNFs can be furthermore classified by the glass transition temperatures, T_g, due to it about 83～85 ℃ for CNF-1 and CNF-3 and about 99～100 ℃ for CNF-2 and CNF-4 (Figure 7-69). This finding suggested that the fabrication of lignin-based CNFs with small diameter is controlled by the carbonization stage and the temperature is a key factor because it low or high would induce lower or higher T_g. In other words, this means that the fiber shape and size are controlled by the T_g of CNF and the thermal conditions would control this parameter. On the basis of above discussion, therefore, the mechanism on formation of alkali lignin-based CNF is understood that the enhancement of the carbonization temperature would enhance the T_g of sample to benefit the non-carbon components gasified and meanwhile the carbon component recrystallized. This is reasonable because the high temperature can fast break the bonds between the carbon and non-carbon components to help the non-carbon components soon gasified and the residue carbons recrystallized.

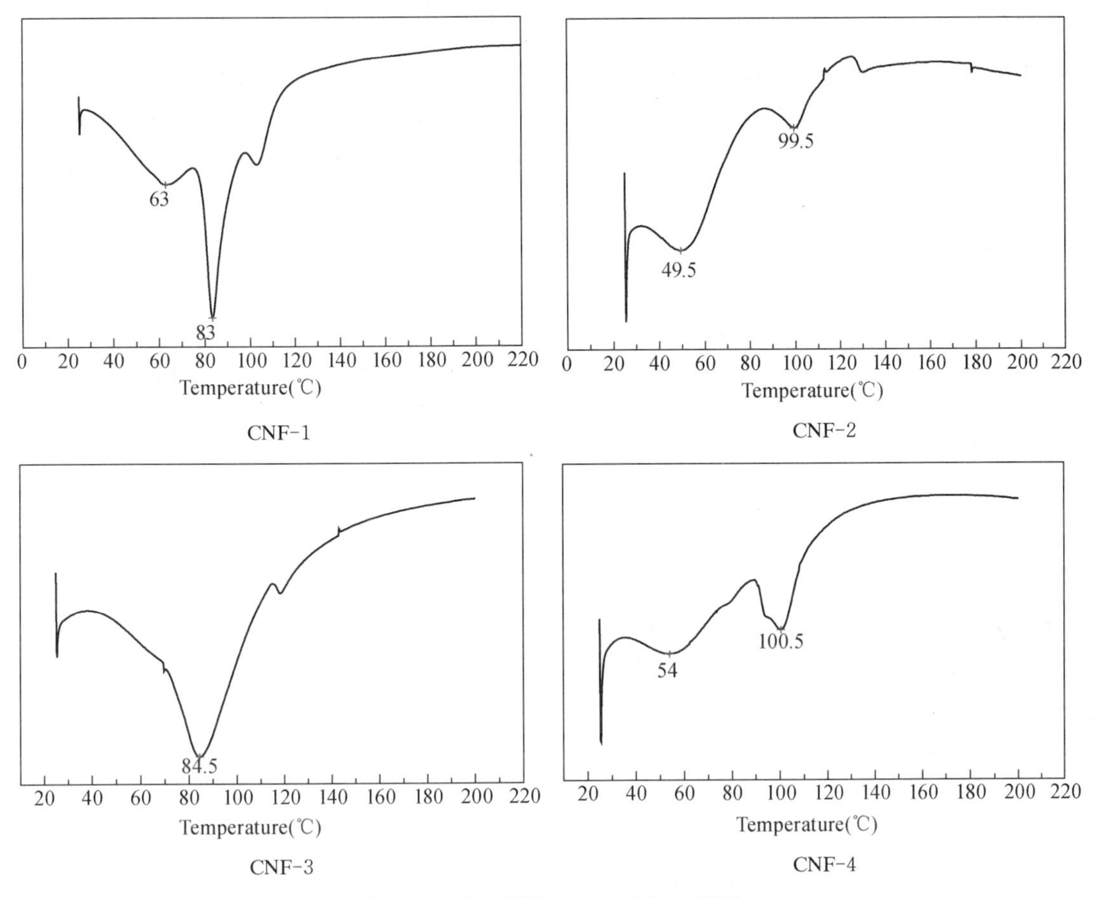

Figure 7-69 DSC curves of four CNFs

7.11.2 Formation of lignin-based activated carbon fiber

As one kind of carbon materials, the activated carbon fibers (ACFs) have been broadly studied and applied. Noted to control, the pore structure is necessary and required for the fabrication of ACFs. Considering that the phenolic resin has been broadly applied to prepare ACFs, and the lignin is one of the most abundant biomacromolecules existing in the plant kingdom, which has the phenyl-propane structure similar to the phenol-formaldehyde resin, recently, we therefore took lignin to replace the phenol/formaldehyde to prepare a phenolic resin and then furthermore to apply this resin to fabricate lignin-based carbon films.

Because the lignin was found a dominator of the carbon film structure, it is furthermore aimed to apply lignin to fabricate ACFs with controllable structure and properties.

Applying the powder alkali lignin with molecular weight M_w of about 28 000 and M_n of 5 000, the phenol (99.5%), formaldehyde (37%), and sodium hydroxide (97%) all in analytic grade as raw materials, the synthesis of lignin-phenol-formaldehyde (LPF) resin was initially carried out in the presence of 0.1M NaOH. To mix the phenol and formaldehyde with a molar ratio of 6 : 7, this hybrid was charged into a reactor, and a NaOH solution was then added to keep the mixture under an alkaline condition to start the reaction. The lignin powder was subsequently added with different contents, for example, 8 wt%, 14 wt%, and 20 wt%, respectively. The mixture was then heated to about 95 ℃ and kept until the resol resin was formed. After cooling, the LPF resin fibers were spun via wet spinning using HCl/formaldehyde (1 : 1) solution as coagulator. The thermal treatment was performed as two steps, the first is to oxide the as-spun fibers (ASFs) in air atmosphere by heating the temperature from 25 ℃ to 250 ℃ at a rate of 2 ℃/min and kept for 1 h. The second is performed under nitrogen atmosphere by heating the temperature from 250 ℃ to 800 ℃ at a rate of 180 ℃/h and kept for another 1 h.

The FTIR spectra of lignin-phenol-formaldehyde (LPF) resin, neat resin, and original lignin were presented and compared in Figure 7-70. Observe the most characteristic infrared bands of lignin located at about 1 500 cm^{-1} and 1 600 cm^{-1} representing the C ═ C aromatic skeletal vibrations. At 3 400 cm^{-1}, the stretching vibration appears due to the presence of —OH. The peaks located at 1 260 cm^{-1} and 876 cm^{-1} are two typical adsorption peaks contributed by the guaiacyl groups of lignin. The 1 100 cm^{-1} is the C—H vibration of the syringyl group in lignin. In this synthesis process, the formaldehyde caused an electrophilic addition reaction toward the active positions of benzene rings, and the addition reaction occurred between the formaldehyde and C_5 position of benzene ring in the guaiacyl group of lignin. Because both the C_3 and C_5 positions in relation to the syringyl groups they connected with the methoxy groups, respectively, while hard to react with formaldehyde, this leads the guaiacyl groups of lignin of importance to react with formaldehyde as Figure 7-71 described.

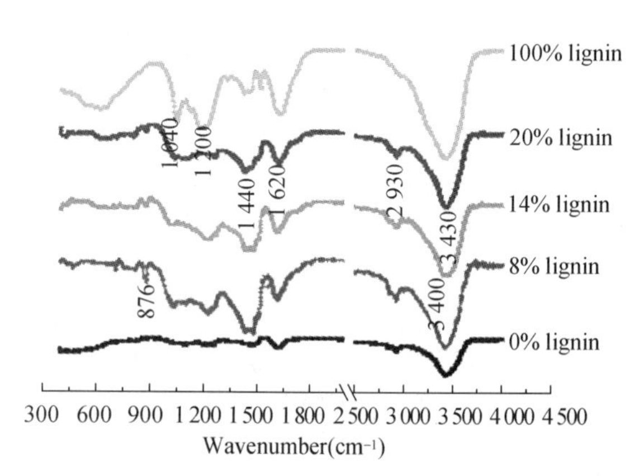

Figure 7-70 FTIR spectra of lignin, LPF resin, and neat resin

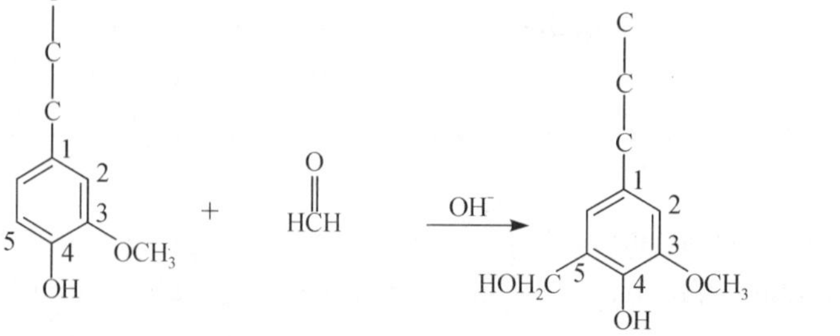

Figure 7-71 Addition reaction occurred between formaldehyde and the C5 position of benzene ring in guaiacyl groups of lignin

A comparison of four LPF resins with different lignin contents found that these samples have the same characteristic peaks suggesting the formation of same chemical structure. This means that the variety of lignin content does not affect the chemical structure of LPF resins. However, it should be addressed, because the increasing of lignin content has been found to cause the increase in intensities for some peaks located between 3 420 cm^{-1} and 3 430 cm^{-1} and the decrease in intensities for peaks located at 2 920 cm^{-1} and 1 470 cm^{-1} representing the vibration of C—H in CH_3 and CH_2, respectively. Additionally, the similar decrease in intensity was also found for the peak located at 876 cm^{-1} representing the guaiacyl groups of lignin. The decrease of the C—H intensity suggested that the formaldehyde caused electrophilic addition that lead the active H reacted on the guaiacyl groups of lignin. In other words, this means that the activity of —OH would be strongly to benefit the AOH groups to react with other groups.

The main intense peaks presented in Figure 7-70 were assigned and summarized in Table 7-24.

The weight losses of these prepared LPF resins traced by TG were described by recorded TG/DTG curves and showed in Figure 7-72, where each sample corresponding to one lignin content varied from 0%, 8%, 14% to 20%, respectively.

Table 7-24 IR peaks assignments for lignin and LPF resin

Wavenumber (cm^{-1})	Assignments[a]
876	γ(C—H); C_2, C_5, C_6 in guaiacyl group
1 020~1 030	s(C—O); aliphatic —OH and —O—
1 100	s(C—H); syringyl group
1 210~1 220	s(C—O); phenolic hydroxyl and ether group
1 260	s(C—O); guaiacyl group
1 430	s(C—C); aromatic ring skeleton
1 470~1 480	δ_a(CH, CH_2)
1 516	γ (aromatic skeleton)
1 620	s(C=C); aromatic
2 920	s(C—H); saturated
3 420	s(phenolic and aliphatic —OH)

As known if there is only a single degradation process, the simple methods have been implemented to treat the thermal degradation of materials. However, when the decomposition occurs through different processes, the analysis is not so simple, especially when these processes may appear in overlapping temperature ranges. According to characteristics values of these TG/DTG curves (Figure 7-72), the degradation processes for these resins have three stages. In the first stage, which related to temperature below 300 ℃, a relatively low percentage of the weight was lost mainly due to the release of excess phenol, aldehyde, absorbed water, and the water of by-product due to the continue synthesis of resin from resol to resitol and to resite. The second stage was occurred within the range of 300～650 ℃ that results in the formation of some products such as the CO, CO_2, benzaldehyde, phenol, and its polymers with random chain scission and the initial formation of char, and, in this temperature range, most of the polymers decomposition took place. In the third stage, when the temperatures above 650 ℃, the dehydration greatly occurred, and a carbon-link structure (char) was thus gradually formed, generating carbon monoxide as by-products.

Characteristic values obtained from the TG and DTG curves of four different lignin content samples (Figure 7-72) are summarized in Table 7-25, where the TD is the temperature at which the maximum rate of degradation occurs, and obtained from the peak of DTG, namely as (dTG/dT). The M_f is the final weight rate of the sample. T_{ONSET} is defined as the temperature at which the decomposition initiates.

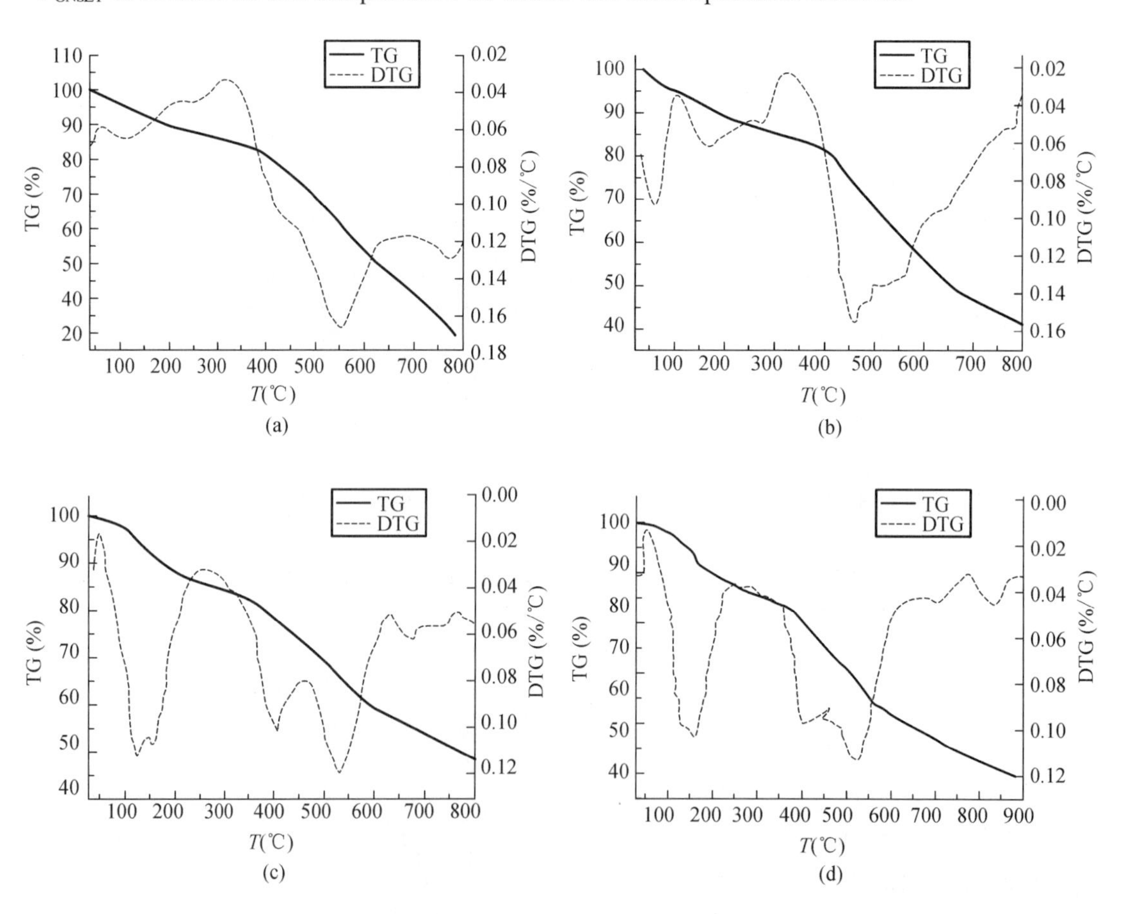

Figure 7-72 TG/DTG curves of ACFs with different lignin contents

According to Table 7-25 and to compare with the neat resin, the temperature ranges of step II of resins, which contained lignin content become small, the T_{ONSET} is shifting to a higher temperature, although the change is tiny only 5～8 ℃. A further comparison of these samples found that the one with higher content of lignin has a higher temperature range in step II and a higher final weight lose rate. This thermobehavior is considerable, because lignin is a complex, heterogeneous, and three-dimensional polymer, which was formed from the enzyme-initiated dehydrogenative free-radical polymerization of cinnamyl alcohol-based precursors, and this leads some small molecules such as phenol and aldehyde to come into the interior of lignin.

Table 7-25 Thermal properties of ACFs

(Thermal degradation values of the LPF resins during thermal treatments)

Sample	Curve	Step Ⅰ	Step Ⅱ	Step Ⅲ	M_f(%) 800 ℃
1. Neat resin	Temp. range (℃)	53~257	257~643	643~800	29.45
	Weight loss (%)	12.61	39.93	19.02	
	TD (℃)	111	553	—	
2. Lignin content 8% resin	Temp. range (℃)	38~266	266~555	555~800	41.02
	Weight loss (%)	13.76	28.61	19.98	
	TD (℃)	63 172	461	—	
3. Lignin content 14% resin	Temp. range (℃)	44~264	264~631	631~800	47.56
	Weight loss (%)	14.53	28.37	9.74	
	TD (℃)	120 159	406 534	—	
4. Lignin content 20% resin	Temp. range (℃)	53~260	260~641	641~800	52.63
	Weight loss (%)	13.03	27.53	6.54	
	TD (℃)	153	405 524	—	

Besides, lignin has the aldehyde and phenolic hydroxyl groups; this benefits to react with the phenol/aldehyde effectively. During the carbonization process, two kinds of carbon structures were formed, one is the "hard carbon" formed from phenol and formaldehyde, and another is the "soft carbon" formed from lignin. Because of the threedimensional structure of lignin, the hollow "soft carbon" is reasonably filled with "hard carbon" to prevent the latter easily degradation. A relative scheme was described in Figure 7-73.

The results of EA on the ASFs, preoxidized fibers, POFs, and ACFs with 20% lignin content were summarized in Table 7-25. A comparison of these values found that the ACFs have a higher carbon content than that of ASFs indicating the ACFs structured by carbon. The degree of preoxidation is found to cause a higher oxygen percentage to imply the process of this thermal stage that leads the formation of more stable ring structures between the benzene rings by the C—O groups. These structures may benefit the fiber maintaining its form to reduce the influence from the melting deformation in the next thermal stage. ACFs have been also found with slightly higher nitrogen content, and this was perhaps caused by carbonization related nitrogen atmosphere due

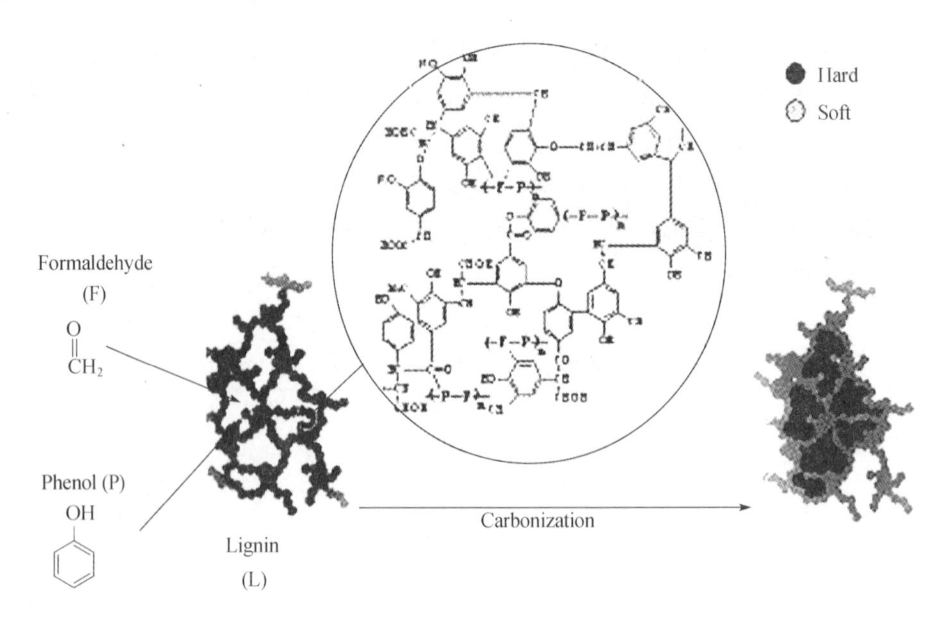

Figure 7-73 Schematic representation of the PFL resin structure

to little nitrogen reacted with LPF resin to increase the nitrogen content.

The SEM photographs of CF samples with lignin contents varied from 8%, 14% to 20% are showned in Figure 7-74, respectively. A comparison of the SEM photographs

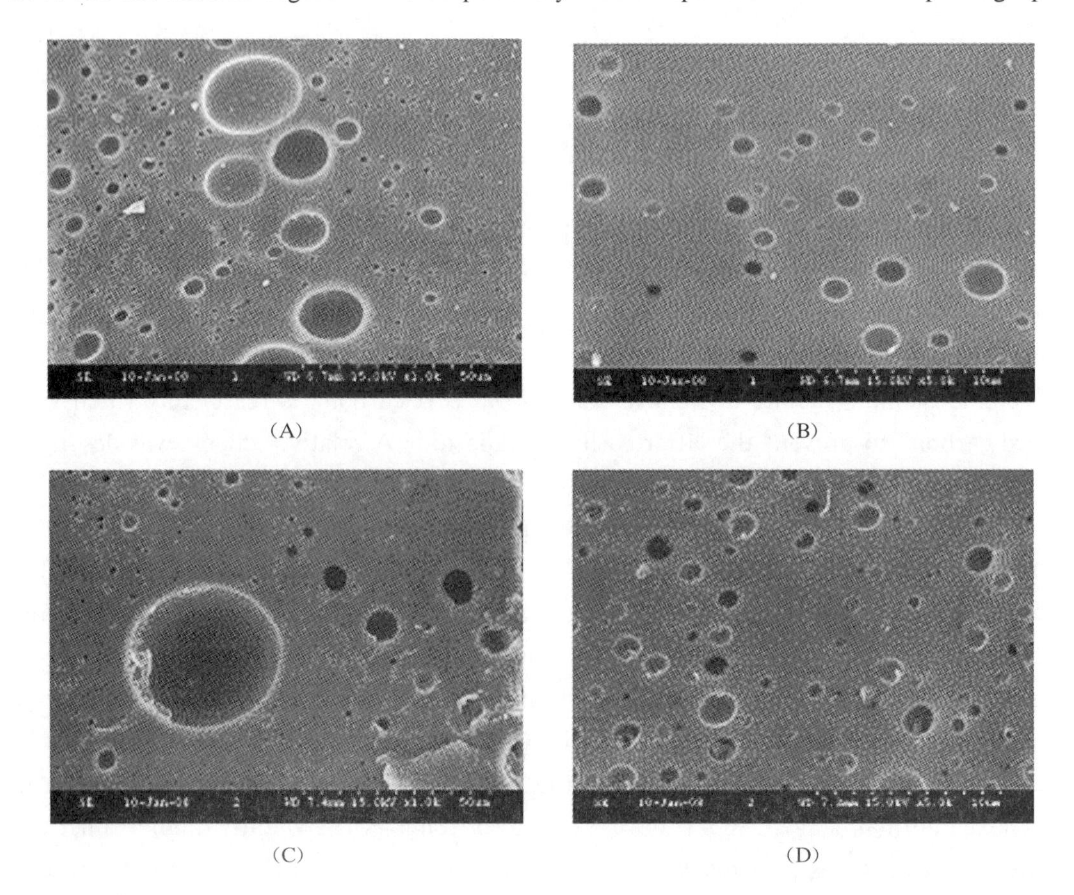

(A) (B) (C) (D)

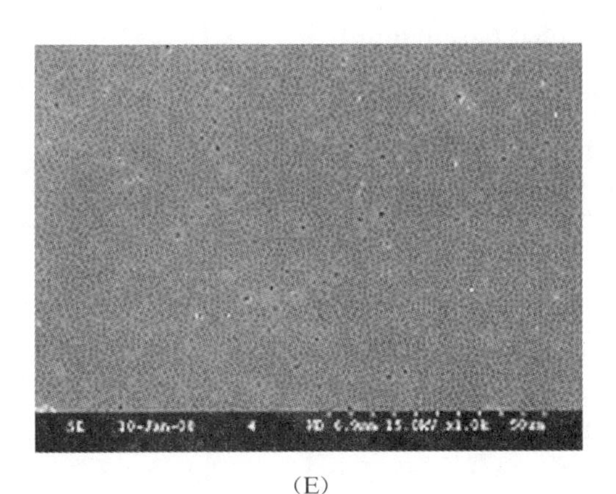

(E)

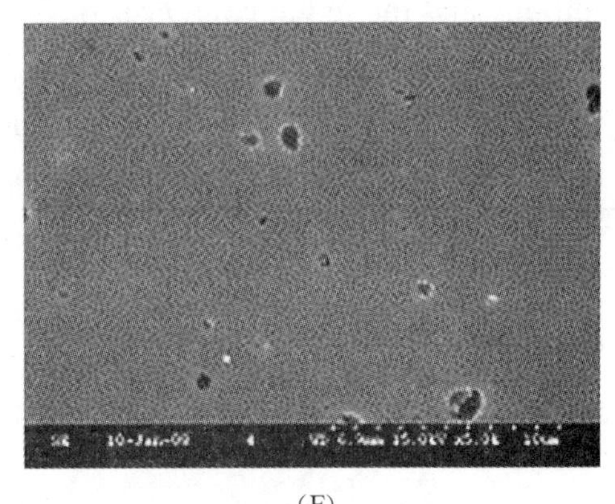

(F)

Figure 7-74 SEM photographs of ACFs with 8% lignin content (A) enlarged by 1 000 times; (B) enlarged by 5 000 times, with 14% lignin content; (C) enlarged by 1 000 times; (D) enlarged by 5 000 times, with 20% lignin content; (E) enlarged by 1 000 times; and (F) enlarged by 5 000 times

on these ACFs with lignin content at 8% and 14% found that their pores presented in a very wide range, for example, the bigger ranged from 5 μm to 15 μm and the smaller ranged from 0.5 μm to 1.5 μm. Noted Figure 7-74 (E~F) showed an obviously different phenomenal for ACFs with lignin content at 20% when compared with other two ACFs as seen in Figure 7-74(A~D), because all pores of the 20% lignin content-based ACFs are observed in the range of 100~800 nm and the pore size distribution in narrow. To compare this work obtained ACFs with literature reported, ACFs found that there are two differences between these ACFs, the first is that the traditional ACFs were prepared by steam activation, and before steam activation, the fibers have been carbonized. The chemical activity of the surface of the fibers was thus very low, and the chemical etching was hard to carry out by steam and CO_2. Additionally, because the graphite-like structure was formed in ACFs, the pores were hard to extend. This means that the traditional methods were hard to get macrospore ACFs. Although in this work, there was not an individual step of activation, because the whole activation process went along with the carbonization process. Therefore, small molecules such as the steam water and formaldehyde would be released during the preoxidation stage. At that time, the precursor fibers have been not formed the graphite-like structure, so that the gas can distract the macromolecular chain segment to form macrospores. The second difference is, in this work, the lignin was directly used as a raw material and its three-dimensional structure would restrict the oversize pore formation and to play a role as pore size controller.

The adsorption data of ACFs were presented in Table 7-26 where the comparison of these data indicates that the 14% lignin content-resulted ACFs have the max porosity and UV adsorption. This suggests that the ACFs could be fabricated under optimized condition, and the lignin content is of importance parameter.

Table 7-26 Elemental analysis of ACF
(Elemental analyses of as-spun fibers, preoxidized fibers, and activated carbon fibers, with 20% lignin content)

Samples	C(%)	H(%)	N(%)	O(%)
ASFs	67.25	5.774	<0.100	26.876
POFs	59.96	3.008	<0.100	36.932
ACFs	91.88	0.822	0.290	7.008

Recommending reading

[1] Z. X. Wang, Q. Shen, Q. F. Gu. *Carbohydrate Polymer*, **2004**, 57, 415-418.
[2] Q. Shen, H. G. Ding, L. Zhong. *Colloids and Surfaces B*, **2004**, 37, 133-136.
[3] Q. Shen, Q. F. Gu, J. F. Hu, X. R. Teng, Y. F. Zhu. *J. Colloids and Interface Sci.*, **2003**, 267, 333-336.
[4] Z. Z. Shao, F. Vollrath. *Nature* **2002**, 418, 741.
[5] H. J. Jin, S.V. Fridrikh, G.C. Rutledge, D.L. Kaplan. *Biomacromolecules*, **2002**, 3, 1233.
[6] U. J. Kim, J. Park, C. Li, H.J. Jin, R. Valluzzi, D. L. Kaplan. *Biomacromolecules*, **2004**, 5, 786.
[7] X. Wang, H. J. Kim, P. Xu, A. Matsumoto, D. L. Kaplan. *Langmuir*, **2005**, 21, 11335.
[8] F. G. Omenetto, D. L. Kaplan. *Nat. Photonics*, **2008**, 2, 641.
[9] S. Ghosh, S. T. Parker, X. Wang, D. L. Kaplan, J. A. Lewis. *Adv. Funct. Mater.*, **2008**, 18, 1883.
[10] D. H. Kim, Y. S. Kim, J. Amsden, B. Panilaitis, D. L. Kaplan, F. G. Omenetto, M. R. Zakin, J. A. Rogers. *Appl. Phys. Lett.*, **2009**, 95.
[11] J. Yuan, Y. Xu, A. H. E. Muller. *Chem. Soc. Rev.*, **2011**, 40, 640-655.
[12] M. Colombo, S. Carregal-Romero, M. F. Casula, L. Gutierrez, M. P. Morales, I. B. Bohm, J. T. Heverhagen, D. Prosperi, W. J. Parak. *Chem. Soc. Rev.*, **2012**, 41, 4306.
[13] J. Thevenot, H. Oliveira, O. Sandre, S. Lecommandoux. *Chem. Soc. Rev.*, **2013**, 42, 7099-7116.
[14] P. Rioux, S. Ricard, R. H. Marchessault. *J. Pulp Pap. Sci.*, **1992**, 18, 39.
[15] L. Raymond, J. F. Revol, D. H. Ryan, R. H. Marchessault. *Chem. Mater.*, **1994**, 6, 249.
[16] G. F. Yin, Z. B. Huang, M. Deng, J. W. Zeng, J. W. Gu. *J. Mater. Chem.*, **2008**, 18, 28.
[17] Y. Lu, L. Dong, L.C. Zhang, Y. D. Su, S. H. Yu. *Nano Today*, **2012**, 7, 297.
[18] J. A. Carrazana-Garcıa, M.A. Lopez-Quintela, J. Rivas-Rey. *Colloid Surf. A*, **1997**, 121, 61.
[19] R. H. Marchessault, P. Rioux, L. Raymond. *Polymer*, **1992**, 33, 4024.
[20] S. Zakaria, B. H. Ong, T. G. M. van de Ven. *Colloid Surf. A*, **2004**, 251, 1.
[21] S. Zakaria, B. H. Ong, S. H. Ahmad, M. Abdullah, T. Yamauchi. *Mater. Chem. Phys.*, **2005**, 89, 216.

[22] E. Sourty, D. H. Ryan, R. H. Marchessault. *Chem. Mater.*, **1998**, 10, 1755.

[23] M. Rubacha. *J. Appl. Polym. Sci.*, **2006**, 101, 1529.

[24] R. P. Swatloski, J. D. Holbrey, J. L. Weston, R. D. Rogers. *Chem. Today*, **2006**, 24, 31.

[25] N. Sun, R. P. Swatloski, M. L. Maxim, M. Rahman, A. G. Harland, A. Haque, S. K. Spear, D. T. Daly, R. D. Rogers. *J. Mater. Chem.*, **2008**, 18, 249.

[26] J. Q. Dong, Q. Shen. *J. Polym. Sci. B*, **2009**, 47, 2036.

[27] J. Q. Dong, Q. Shen. *J. Appl. Polym. Sci.*, **2012**, 126:S1, E10-E16.

[28] Q. Yang, F. D. Dou, B. R. Liang, Q. Shen. *Carbohydr. Polym.*, **2005**, 59, 205.

[29] Q. Yang, F. D. Dou, B. R. Liang, Q. Shen. *Carbohydr. Polym.*, **2005**, 61, 393.

[30] Z. J. Gu, J. T. Wang, L. L. Li, L. F. Chen, Q. Shen. *Mater Lett.*, **2014**, 117, 66-68.

[31] Z. J. Gu, J. R. Ye, W. Song, Q. Shen. *Mater Lett.*, **2014**, 121, 12-14.

[32] Z. J. Gu, Q. C. Zhang, Q. Shen. *J. Polym. Res.*, **2015**, 22:2, 1-4.

[33] J. T. Wang, L. L. Li, Q. Shen. *Int'l J. Biological Macromolecules*, **2014**, 63, 205-209.

[34] J. T. Wang, L. L. Li, L. H. Jiang, Q. Shen. *Mater Sci Eng C.*, **2014**, 34, 417-421.

[35] J. R. Ye, S. Zhai, Z. J. Gu, N. Wang, H. Wang, Q. Shen. *Mater Lett.*, **2014**, 132, 377-379.

[36] N. Wang, H. Li, T. Y. Chen, J. T. Wang, Q. Shen. *Mater Lett.*, **2014**, 137, 203-205.

[37] Q. Shen, T. Zhang, W. X. Zhang, S. Chen, M. Mezgebe. *J. Appl Polym Sci.*, **2011**, 121:2, 989-994.

Problems

1. Please design a new method for directly obtaining functional silk fiber.
2. Please find different interfacial polymerization processes and do a comparison.